T0206000

SpringerBriefs in Molecular Science

SpringerBriefs in Molecular Science present concise summaries of cutting-edge research and practical applications across a wide spectrum of fields centered around chemistry. Featuring compact volumes of 50 to 125 pages, the series covers a range of content from professional to academic. Typical topics might include:

- A timely report of state-of-the-art analytical techniques
- A bridge between new research results, as published in journal articles, and a contextual literature review
- A snapshot of a hot or emerging topic
- An in-depth case study
- A presentation of core concepts that students must understand in order to make independent contributions

Briefs allow authors to present their ideas and readers to absorb them with minimal time investment. Briefs will be published as part of Springer's eBook collection, with millions of users worldwide. In addition, Briefs will be available for individual print and electronic purchase. Briefs are characterized by fast, global electronic dissemination, standard publishing contracts, easy-to-use manuscript preparation and formatting guidelines, and expedited production schedules. Both solicited and unsolicited manuscripts are considered for publication in this series.

More information about this series at http://www.springer.com/series/8898

Yu Lan · Cheng-Xing Cui · Song Liu ·
Chun-Xiang Li · Ruopeng Bai

Computational Advances of Rh-Catalyzed C–H Functionalization

From Elementary Reaction to Mechanism

 Springer

Yu Lan
Green Catalysis Center
College of Chemistry
Zhengzhou University
Zhengzhou, Henan, China

Song Liu
Chongqing Key Laboratory
of Environmental Materials
and Remediation Technologies
Chongqing University of Arts and Sciences
Chongqing, China

Ruopeng Bai
Chongqing Key Laboratory of Theoretical
and Computational Chemistry
School of Chemistry and Chemical
Engineering
Chongqing University
Chongqing, China

Cheng-Xing Cui
School of Chemistry and Chemical
Engineering
Henan Institute of Science and Technology
Xinxiang, Henan, China

Chun-Xiang Li
School of Chemistry and Chemical
Engineering
Henan Institute of Science and Technology
Xinxiang, Henan, China

ISSN 2191-5407   ISSN 2191-5415 (electronic)
SpringerBriefs in Molecular Science
ISBN 978-981-16-0431-7   ISBN 978-981-16-0432-4 (eBook)
https://doi.org/10.1007/978-981-16-0432-4

This Springer imprint is published by the registered company Springer Nature Singapore Pte Ltd.
The registered company address is: 152 Beach Road, #21-01/04 Gateway East, Singapore 189721, Singapore

Preface

Selective formation of carbon–carbon or carbon–heteroatom bonds through transition metal-catalyzed direct functionalization of unreactive C–H bonds has attracted considerable attention in chemical research and industry with regard to atom and step economy. Transition metal catalysts, including Pd, Rh, Ru, Co, Ir, Ni, and Pt catalysts, are crucial for C–H bond cleavage and further transformations. In particular, Rh, which might be the best alternative to Pd, shows significant catalytic activity in C–H functionalization and has attracted wide attention in synthetic chemistry. This has become an increasingly important strategy for the construction of complex organic molecules.

In recent years, a broad range of experimental transformations can be achieved through Rh-catalyzed C–H bond functionalization. Today, the goal for Rh-catalyzed C–H activation is to selectively activate targeted C–H bonds under mild conditions to synthesize specific organic molecules in an efficient, highly selective, and green manner. Along with the huge progress that has been witnessed experimentally, a complete understanding of the mechanism of a given reaction can lead to improved reactions and enable the development of novel reactions. Recently, the computational study of Rh-catalyzed C–H functionalization has achieved significant accomplishments. Both the detailed reaction mechanism and origin of selectivity in Rh-catalyzed C–H functionalization have been better elucidated by the computational study.

This volume focuses on the theoretical aspects of Rh-catalyzed C–H bond functionalization, providing the first comprehensive and systematic summary of theoretical advances in Rh-catalyzed C–H functionalization in recent decades. We present a general view of organometallic chemistry and a brief history of Rh-catalyzed C–H functionalization. We point out the importance of using the computational tool to study the mechanism of Rh-catalyzed C–H functionalization. In addition, this volume presents the computational methods in theoretical studies of Rh-catalyzed C–H functionalization. We also summarize the elementary reactions in Rh-catalyzed C–H functionalization from the aspect of nucleophilic C–H bond activation and electrophilic C–H bond activation. And then, the C–C and C–X bond formation involving arylation, alkylation, alkenylation, alkynylation, carbonylation, hydroacylation, and annulation reactions catalyzed by Rh are discussed. In each section, the order of the elementary reactions, changes in the Rh oxidation state, and their relationship with the C–H bond activation mode are emphasized to demonstrate mechanistic diversity.

The wide range of theoretical advances presented here would provide deeper mechanistic understandings and offer guidance for optimizing or designing ligands and reactions on the concepts of Rh-catalyzed C–H functionalization. This summary of mechanistic studies on Rh-catalyzed C–H bond activation and functionalization will promote the development of transition metal catalysis and lead to the discovery of new concepts in this area.

We are grateful to all the chapter authors, who have contributed to create this multiple-facet volume. We dedicate this volume to all chemists and students who are contributing, via C–H bond activation and functionalization, to discover safe catalytic transformations in the organic chemistry field.

Zhengzhou, China   Yu Lan
Xinxiang, China   Cheng-Xing Cui
Chongqing, China   Song Liu
Xinxiang, China   Chun-Xiang Li
Chongqing, China   Ruopeng Bai

Contents

Chapter 1
Theoretical View of Rh-Catalyzed C–H Functionalization

Yu Lan

1.1 A Brief History of Rh-Catalyzed C–H Functionalization

1.1.1 General View of Organometallic Chemistry

The synthetic organic chemistry usually focuses on "carbon" to spread related research, which could afford various strategies for the building of molecular framework, functional group transformations, and controlling stereochemistry in more sophisticated molecules [1–13]. Therefore, selective formation of the new covalent bond between the carbon atom and some other atoms, involving nitrogen, oxygen, sulfur, halogen, boron, and phosphorus, becomes one of the most important aims for synthetic organic chemistry [14, 15]. The coupling reactions provide practical and efficient methods for the formation of new carbon–carbon and carbon–heteroatom bonds [16–25]. Generally, cross-coupling reactions build covalent bonds between nucleophiles and electrophiles [26–33]. Reductive coupling combines two electrophiles by gaining electrons from reductive agents [34–40]. When oxidations are involved, two nucleophiles would conduct oxidative coupling to lose electrons and form carbon–carbon and carbon–heteroatom bonds [41–47] (Scheme 1.1).

In organic chemistry, the nucleophile is an electron-rich molecule that contains a lone pair of electrons or a polarized bond, the heterolysis of which also could yield a lone pair of elections [48–50] (Scheme 1.2). According to this concept, organometallic compounds, alcohols, halides, amines, and phosphines with a lone pair of electrons are nucleophiles [51–55]. Some nonpolar π bonds including olefins and acetylenes, which could donate the π bonding electrons, are often considered to be nucleophiles [56–58]. Moreover, the C–H bonds of hydrocarbons can be considered to be nucleophiles because the electronegativity of carbon is higher than that of hydrogen, which could deliver a proton to form a formal carbon anion [59–65]. Correspondingly, the electrophile is an electron-deficient molecule that contains unoccupied orbitals or low energy antibonding molecular orbital, which could accept

© The Author(s), under exclusive license to Springer Nature Singapore Pte Ltd. 2021
Y. Lan et al., *Computational Advances of Rh-Catalyzed C–H Functionalization*,
SpringerBriefs in Molecular Science,
https://doi.org/10.1007/978-981-16-0432-4_1

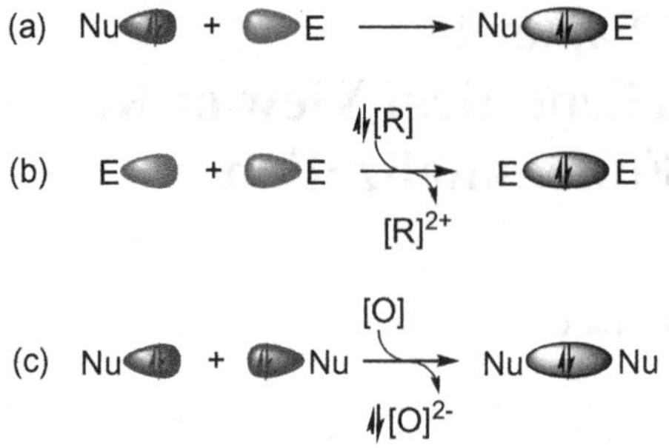

Scheme 1.1
(**a**) Redox-neutral
cross-coupling reactions,
(**b**) reductive cross-coupling
reactions, and (**c**) oxidative
cross-coupling reactions

(a) Nu⬤ + ⬤E ⟶ Nu⬤E

(b) E⬤ + ⬤E $\xrightarrow[\text{[R]}^{2+}]{\text{[R]}}$ E⬤E

(c) Nu⬤ + ⬤Nu $\xrightarrow[\text{[O]}^{2-}]{\text{[O]}}$ Nu⬤Nu

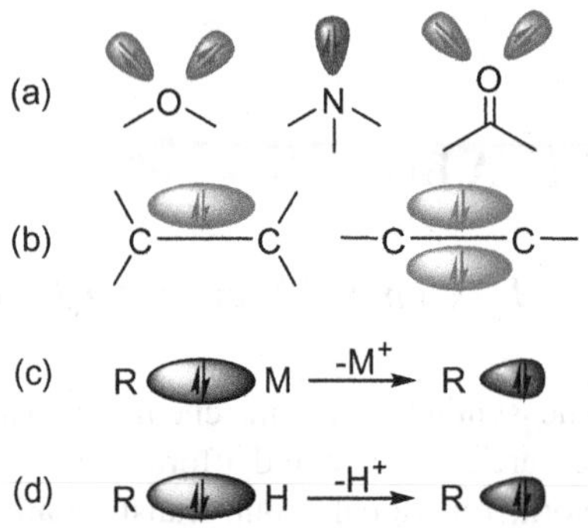

Scheme 1.2 Some selected
examples of nucleophiles:
(**a**) lone-pair of electrons,
(**b**) unsaturated π bonds,
(**c**) organometallic
complexes, and (**d**) C–H
bonds

(a)

(b)

(c) R⬤M $\xrightarrow{-M^+}$ R⬤

(d) R⬤H $\xrightarrow{-H^+}$ R⬤

the electrons from nucleophiles. In the organometallic chemistry, cationic carbons, which usually come from the heterolysis of carbon–halogen bonds, are electrophile. Polar π bonds, including carbonyl compounds and imines, also could be considered to be electrophile, which involves a low energy π antibonding. In addition, Fisher type singlet carbene has an electron pair filling one sp^2 hybrid orbital and an unoccupied p orbital, which could be considered to be either nucleophile or electrophile in coupling reactions.

The reactivity of the direct coupling reactions is relatively low; the introduction of transition metal catalysts provides novel strategies to construct covalent bonds, thus offering a great opportunity to derivative raw chemicals with little functionality to synthetically versatile molecules [66–74]. Though the compounds involving carbon–hydrogen (C–H) bonds could be considered as unique nucleophiles, the reactivity of such compounds is always under restriction. The poor reactivity of C–H bonds could be often attributed to their high bond energies, which results from the high energy of the antibonding orbital and low energy of the bonding orbital for the C–H bonds. However, the use of C–H bonds as a transformable functional group is advantageous because these bonds are typically the most abundant functionality in organic molecules. Direct conversion of these bonds to the desired functionality shortens synthetic pathways, saving reagents, solvents, and labor [75]. Less chemical waste is generated as well. Fortunately, the introduction of transition metal catalysts makes the site-selective cleavage and functionalization of C–H bonds to be feasible [76–78].

In recent decades, the selective formation of carbon–carbon and carbon–heteroatom bonds through transition metal-catalyzed direct functionalization of unreactive carbon–hydrogen bonds has attracted considerable attention in chemical research and industry from the perspective of atom and step economy [79–86]. High valence transition metal can obtain electrons from nucleophile, which leads to the transformation of nucleophile into electrophile. The newly generated electrophile can couple with other nucleophiles to form a covalent bond, which is named oxidative coupling reaction. Meanwhile, the reduced transition metal can be oxidized by an exogenous oxidant for regeneration. Correspondingly, low valence transition metal can donate electrons to electrophile leading to the transformation of electrophile into nucleophile, which can react with another electrophile to form a covalent bond. Accordingly, it is named a reductive coupling reaction. The oxidized transition metal also can be reduced by exogenous reductant.

Transition metal catalysts, especially Pd, Rh, Ru, Ir, Ni, and Cu catalysts, are crucial for C–H bond cleavage and further transformations [87–92]. In this area, extensive efforts have been devoted to the development of Rh catalysts in catalytic conversion of C–H bonds. It has become an increasingly important strategy for the construction of complex organic products. A broad range of transformations could be achieved through rhodium-catalyzed C–H bond functionalization. For example, Ellman and Bergman have developed a series of efficiently and intermolecularly alkylation of nitrogen heterocycles with a wide range of functionalized olefins via Rh-catalyzed C–H bond activation [88, 93, 94]. In addition, an array of Rh(I) and Rh(III)-catalyzed C–H bond functionalization reactions that take advantage of a chelating directing group have also been studied by the same group [95–102]. The use of chelation control in C–H bond functionalization offers several advantages with respect to substrate scope and application to total synthesis [103]. Glorius reported a Cp*Rh(III)-catalyzed reaction of N-phenoxyacetamide with 7-azabenzonorbornadiene leads to a dearomatized spiro structure with a cyclopropane unit [104]. Li and co-workers have also reported a series of C–H activation of arenes using Cp*Rh(III) complexes with high reactivity, stability, and functional group compatibility [105, 106]. In addition, Chang's group has made a contribution to the field of development of Rh-catalyzed direct C–H amination reactions with organic azides [107]. Under their optimized Rh(III)-catalyzed amination conditions, not only sulfonyl azides but also aryl- and alkyl azides could be utilized as facile amino sources in reaction with various types of C(sp^2)–H bonds. In order to realize Rh(III)-catalytic enantioselective transformations, Cramer and co-workers have introduced two families of chiral cyclopentadienyls ligand. The first generation consists of a fused cyclohexane unit having pseudo axial methyl groups as chiral selectors and a rigidifying acetal moiety [108]. The second ligand generation derives from an atop-chiral biaryl-backbone and which possesses adjustable substituents at its 3,3'-positions. Rovis also describes a stunning effect on reactivity and selectivity of Rh-catalyzed C–H bond functionalization reactions by the introduction of achiral mono-, di-, or pentasubstituted cyclopentadienyl ligands [109–111].

As an overview of Rh-catalyzed C–H functionalization, the core is the formation of a rhodium–carbon bond and its further transformation. The rhodium-catalyzed process usually goes through multiple steps as well as complicated catalytic cycles, which originated from the complex bonding pattern of rhodium catalyst and the variation of valance state for the central rhodium element. Consequently, improving the reaction efficiency and yield for rhodium catalysis encountered difficulty. Moreover, the design of catalysis and ligand for the rhodium-catalyzed reaction is still facing both opportunities and challenges. To solve the above-mentioned issues, the understanding of the reaction mechanism is imperative, which could give more information for the detailed reaction process, and help to improve the reaction efficiency and yield.

1.1.2   A Brief History of Rh-Catalyzed C–H Functionalization

Transition metal-catalyzed C–H bond functionalization has been a highly intriguing research topic for the past two decades from the atom- and step-economical points [112–116]. A variety of catalytic processes that use different transition metals and modes for activating the inert C–H bond have been developed. The key issue for the development of C–H bond activation is the selective activation of a targeted C–H bond over the other C–H bonds in the substrate. The most promising and useful strategy is to utilize coordination of a functional group in the substrate to the metal center of a catalyst to conduct the selective activation of C–H bond.

As pioneering work in the Rh-catalyzed selectivity C–H bond activation, Lim and co-workers employed pyridine as a directing group to direct functionalization [117]. The RhCl(PPh$_3$)$_3$ (Wilkinson's catalyst) was used as the active catalyst in this transformation. The isomerization of 1-linear alkyl olefins to 2-olefin makes the coupling reactions give a moderate yield and needs a longer reaction time. The reactions of 3,3-dimethylbutyl-1-ene and vinylsilanes afford high yields and need a lower reaction time because the isomerization cannot conduct (Scheme 1.3).

As relatively stable toward air and moisture, the RhCl(PPh$_3$)$_3$ (Wilkinson's catalyst) was usually used as a catalyst in Rh-catalyzed selectivity C–H bond activation. In 2000, Jun and co-workers [118] reported a Rh-catalyzed *ortho*-alkylation reaction of aromatic imines by using RhCl(PPh$_3$)$_3$ (Wilkinson's catalyst) as the catalyst.

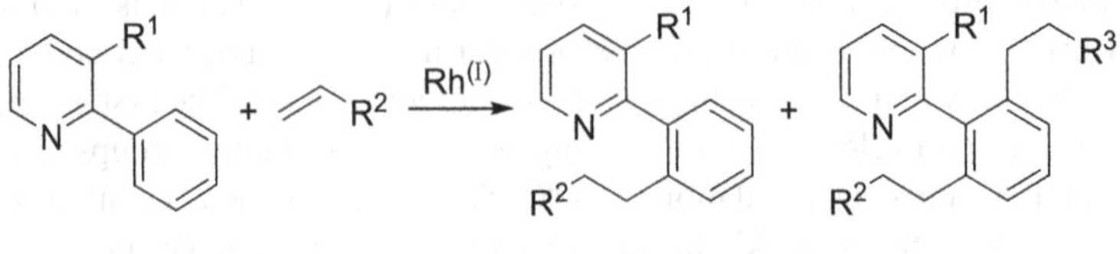

Scheme 1.3   A possible reaction of the *ortho* position alkylation

Under optimized conditions, this reaction was found to have substantially broader scope than the Ru-catalyzed alkylation of aryl ketones (Scheme 1.4).

For the aromatic aldimines as substrates, the Wilkinson's catalyst was ineffective without a co-catalyst [118]. In 2004, Lim and co-workers [119] report the alkylation of aromatic aldimines and ketimines with alkenes by a more electron-donating catalyst system, [RhCl(coe)$_2$]$_2$ (coe = cyclooctene) and tricyclohexylphosphine (Cy$_3$P), without any need for additives. The following hydrolysis and chromatographic isolation of *ortho*-alkylated aryl aldimines afforded the final *ortho*-alkylated aryl aldehydes product. Moreover, the product of overalkylation could be avoided by introducing a substituent at the third position to sterically encumber one site or by blocking one of the *ortho*-positions. Using this strategy, a variety of Rh-catalyzed hydroarylation reactions of olefins have been developed (Scheme 1.5).

When the Rh-catalyzed C–H bond activation and hydroarylation of olefins have seen broad success, only a few examples of the analogous reaction for alkynes have been reported. The Rh-catalyzed dimer- and trimerization reactions of alkynes, in particular terminal alkynes, restrict the application of alkynes in C–H bond functionalization reactions. Moreover, the internal alkynes are often unreactive in Rh-catalyzed C–H bond activation reactions.

The first example of the Rh-catalyzed hydroarylation of alkynes was reported by Lim and co-workers [120]. This reaction proceeded with 2-Phenylpyridine (0.3 mmol) and 2-butyne (2 equivalents) as substrates, RhCl(PPh$_3$)$_3$ (10 mol%) as the catalyst, and in the presence of 10 mol% of PPh$_3$ in toluene. The major double *ortho*-alkenylated product and the mono-alkenylated product were generated at the same time with a ratio of 81:19. When 3 equivalents of 2-butyne and 2-(*p*-tolyl)pyridine were used, the double *ortho*-alkenylated complex could be generated as the unique product in 89% isolated yield. But the scope of alkynes in this reaction was limited primarily to internal and symmetrical alkynes (Scheme 1.6).

Then Jun and co-workers [121] have demonstrated a Rh(I)-catalyzed *ortho*-alkenylation of aromatic ketimines with terminal alkynes using Wilkinson's catalyst. When 1.0 equivalent aromatic ketimines and 1.2 equivalent 1-hexyne or 1-octyne

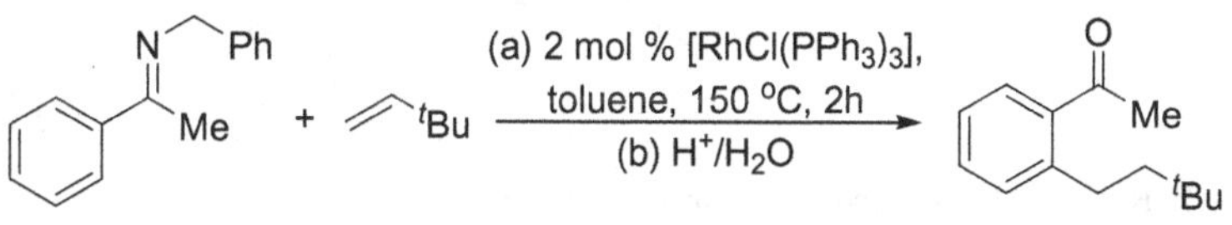

Scheme 1.4 The *ortho*-alkylation reaction of aromatic imines

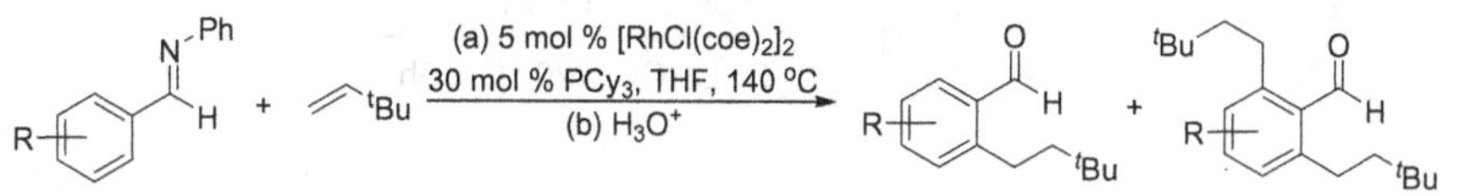

Scheme 1.5 The reaction of the alkylation of aromatic aldimines

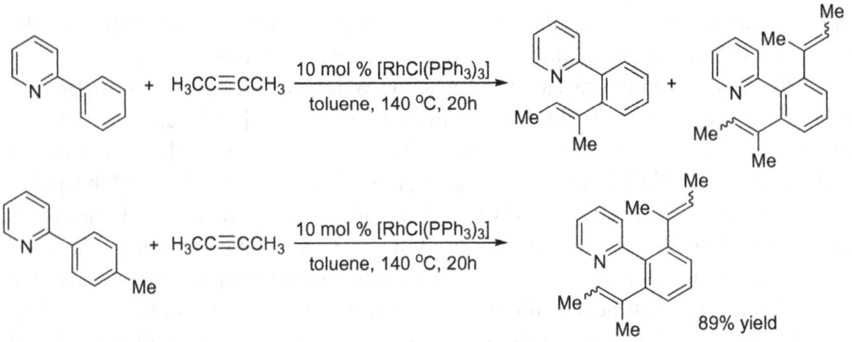

Scheme 1.6 The reaction of 2-phenylpyridine [2-(*p*-tolyl)pyridine] and 2-butyne

were used in this reaction, the monoalkenylated products were generated with a ratio of 100% and good yields (Scheme 1.7).

The Wilkinson's catalyst also can be used for chelation-assisted arylation of 2-arylpyridines. In 1998, Oi and co-workers reported Rh-catalyzed *ortho*-arylation of 2-arylpyridines with arylstannanes [122]. The double *ortho*-arylation could be avoided by blocking one of the ortho-positions on the aryl substituent or by introducing a substituent at the third position of the pyridine ring. Since then a variety of Rh(I)-catalyzed arylation of aromatics have been developed (Scheme 1.8).

Moreover, the earliest Rh-catalyzed directed carbonylation of arenes was reported by Chatani and co-workers in 2004 [123]. It was found that the carbonylation at C–H bonds in N-arylpyrazoles is efficiently catalyzed by $Rh_4(CO)_{12}$ even at 140 °C, provided DMA is used as the solvent. This transformation was also tolerant of electron-donating substituents on the arylpyrazole ring (Scheme 1.9).

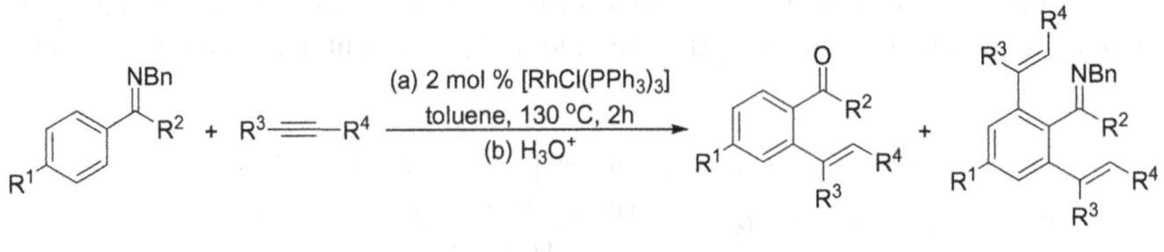

Scheme 1.7 Rh(I)-catalyzed *ortho*-vinylation of ketimines

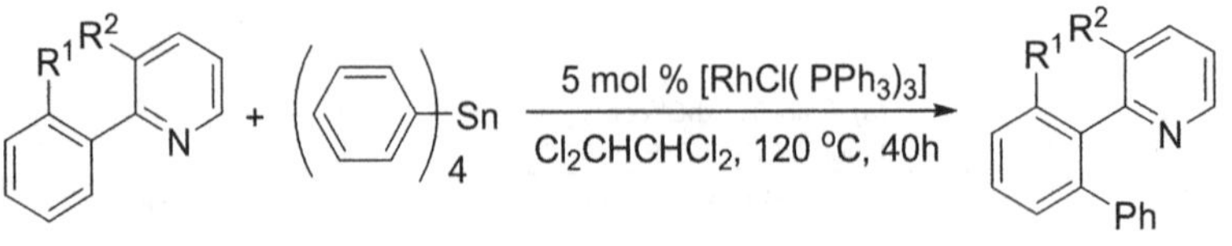

Scheme 1.8 Rh(I)-catalyzed *ortho*-arylation of 2-arylpyridines with arylstannanes

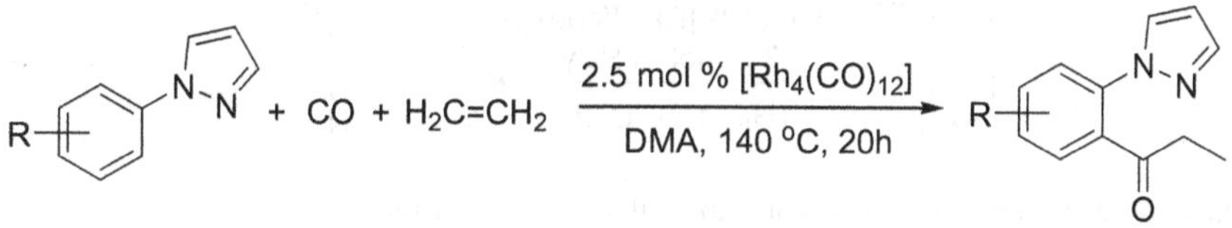

Scheme 1.9 $Rh_4(CO)_{12}$-catalyzed reaction of N-arylpyrazoles with CO and alkenes

The reactivity of the Rh-catalyzed C–H functionalization reactions of olefins is relative lower than that of arenes. In 2002, Jun and co-workers [124] were the first to report on the Rh-catalyzed alkylation of an α,β-unsaturated carbonyl derivative using $RhCl(PPh_3)_3$ as catalyst. The reaction of enone with a variety of alkyl- and silyl-monosubstituted olefins produced a mixture of substituted ketones (Scheme 1.10).

Then Fürstner and co-workers [125] developed a tandem pyridyldirected olefin C–H bond activation and cycloisomerization of a tethered alkylidenecyclopropane using a cationic Rh catalyst generated from $RhCl(PPh_3)_3$ and $AgSbF_6$ in 2007. The reaction proceeded in moderate yields when a rigid tether consisting of an aryl or cyclohexyl ring (Scheme 1.11).

The selective C–H bond functionalization reactions of N-heterocycles represent an appealing approach towards generating a wide variety of substituted heterocyclic complexes. The first example of Rh-catalyzed alkylation of azole by C–H activation was reported by Bergman, Ellman, and co-workers [126]. The Wilkinson's catalyst was an effective catalyst for this transformation. The Rh-precatalyst ($[RhCl(coe)_2]_2$) and phosphine ligands (PCy_3) provided a much more efficient catalyst system and produced a high yield (Scheme 1.12).

The initial investigations into the arylation of heterocycles via Rhodium-catalyzed C–H bond functionalization were also reported by Bergman, Ellman, and co-workers

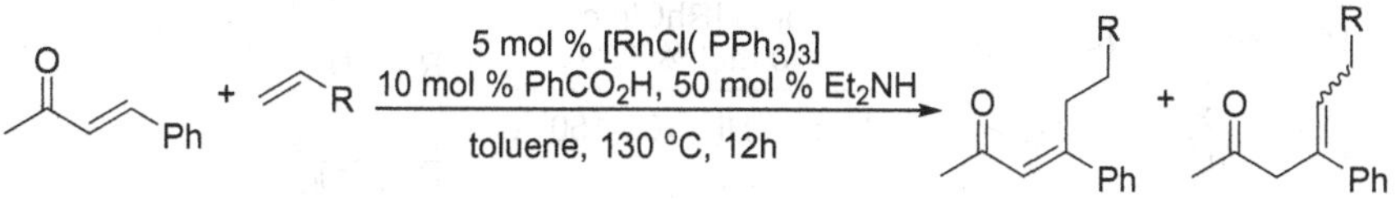

Scheme 1.10 Rh(I)-catalyzed β-alkylation of 4-phenyl-3-buten-2-one with 1-alkene

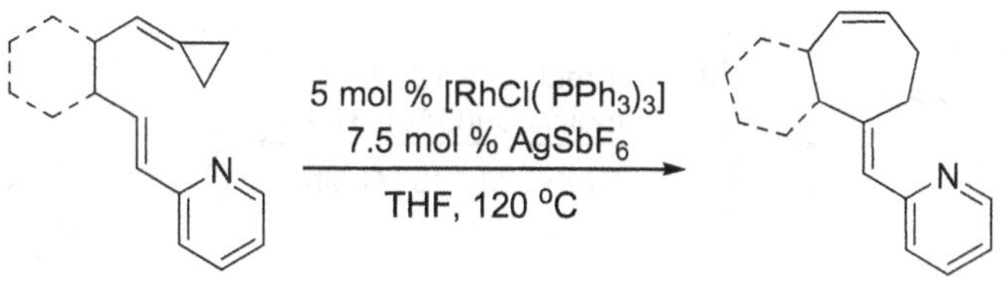

Scheme 1.11 Rh-catalyzed C–H bond activation and cycloisomerization

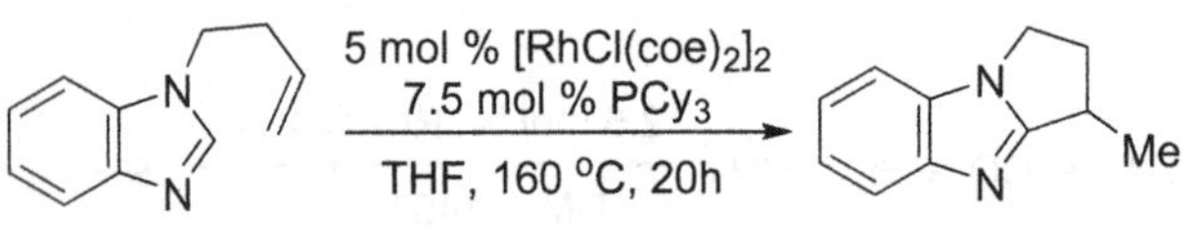

Scheme 1.12 Rh-catalyzed alkylation of azole through C–H activation

[127]. In contrast to the alkylation of azole, a base additive was required in this reaction to achieve good yields (Scheme 1.13).

Despite the abundant application of Wilkinson's catalyst, the Rh(III) catalyst [RhCp*Cl$_2$]$_2$ (Cp* = pentamethylcyclopentadienyl) was also widely used in C–H activations. In 2007, Miura and Satoh [128] reported the first [RhCp*Cl$_2$]$_2$-catalyzed C–H activation of arenes. In this transformation, both electron-rich and electron-deficient benzoic acids could be used as substrates. In addition, this reaction tolerated both alkyl and aryl alkynes as substrates (Scheme 1.14).

Science then, explosive progress has been made in this area, and increasing attention has been devoted to Cp*Rh-catalyzed C–H activation in recent years. Cp*Rh(III) catalysts have stood out with high activity, broad substrate scope, mild conditions, and functional group compatibility. For example, Glorius reported a Cp*Rh(III)-catalyzed reaction of N-phenoxyacetamide with 7-azabenzonorbornadiene leads to a dearomatized spiro structure with a cyclopropane unit [104]. Li and co-workers have also reported a series of C–H activation of arenes using Cp*Rh(III) complexes with high reactivity, stability, and functional group compatibility [105, 106]. In addition, Chang's group has made a contribution to this field of development of Rh-catalyzed direct C–H amination reactions with organic azides [107]. Under their optimized Rh(III)-catalyzed amination conditions, not only sulfonyl azides but also aryl- and alkyl azides could be utilized as facile amino sources in reaction with various types

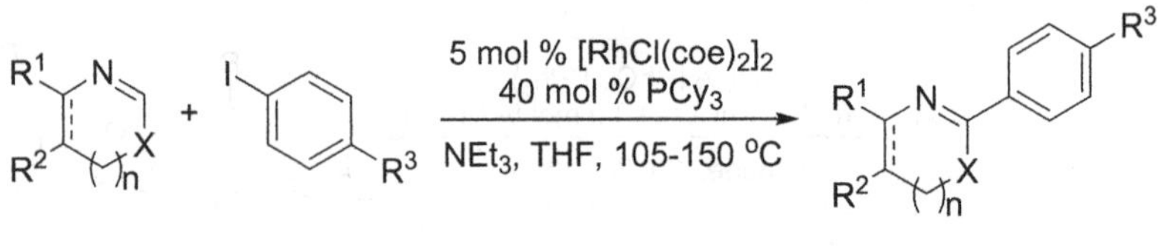

Scheme 1.13 Rh-catalyzed arylation of heterocycles through C–H bond functionalization

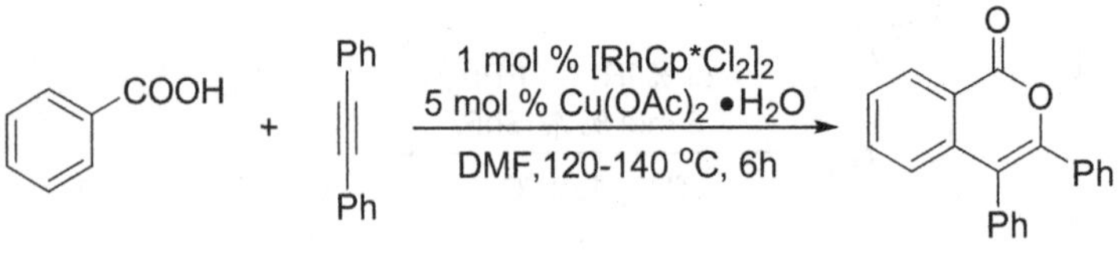

Scheme 1.14 Rh-catalyzed dehydrogenative coupling of benzoic acid with diphenylacetylene

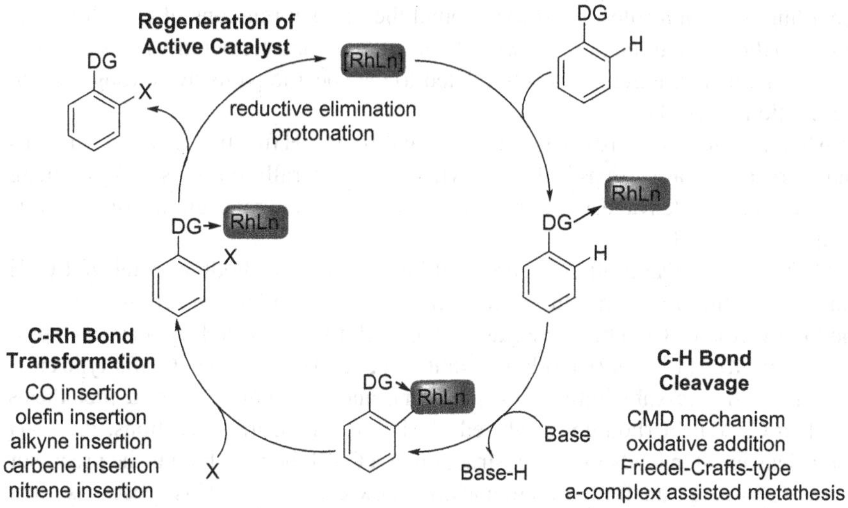

Scheme 1.15   The possible mechanism of Rh-catalyzed C–H functionalization

of C(sp^2)–H bonds. In order to realize Rh(III)-catalytic enantioselective transformations, Cramer and co-workers have introduced two families chiral cyclopentadienyls ligand. The first generation consists of a fused cyclohexane unit having pseudo axial methyl groups as chiral selectors and a rigidifying acetal moiety [108]. The second ligand generation derives from an atrop-chiral biaryl-backbone, which possesses adjustable substituents at its 3'-positions. Rovis also describes a stunning effect on reactivity and selectivity of Rh-catalyzed C–H bond functionalization reactions by introduction of achiral mono-, di-, or pentasubstituted cyclopentadienyl ligands [109–111] (Scheme 1.15).

1.2   Using Computational Tool to Study the Mechanism of Rh-Catalyzed C–H Functionalization

1.2.1   Mechanism of Rh-Catalyzed C–H Functionalization

Along with the explosive progress witnessed from an experimental aspect, a complete understanding of the mechanism for a given reaction can lead to improved reactions and enable the development of novel reactions [129, 130]. Therefore, it is crucial to understand the mechanisms of rhodium-catalyzed C–H bond functionalization by detailed experimental and computational studies.

Generally, reaction mechanism could be considered to be all elementary reactions used to describe a chemical change passing in a reaction [131, 132]. It is to decompose a complex reaction into several elementary reactions and then combine them according to certain rules, so as to expound the internal relations of complex reactions and the internal relations between total reactions and elementary reactions. The rate of chemical reaction is closely related to the specific pathways through which the reaction takes place.

Although the mechanism details for Rh-catalyzed C–H functionalization reactions may vary from case to case, the catalytic cycle generally consists of three main steps: C–H bond cleavage, C–Rh bond transformation, and regeneration of the active catalyst [133–140].

C–H bond cleavage often occurs as the initial step in rhodium-catalyzed C–H functionalization reactions. In general, there are four possible reaction modes for rhodium-mediated C–H bond cleavage (Scheme 1.15): concerted metalation–deprotonation (CMD) [141, 142], oxidative addition (OA) [143], Friedel–Crafts-type electrophilic aromatic substitution [144] (SEAr), and σ-complex assisted metathesis [145] (σ-CAM). In rhodium-catalyzed C–H functionalization reactions, the C–H bond cleavage step leads to the construction of a C–Rh bond, which is the precursor for the following transformation to construct new C–X bonds. This process is often realized through insertion reactions including CO [123, 146, 147], olefin [98, 148–150], acetylene [106, 151, 152], and carbene/nitrene [73, 153, 154] insertion. Formation of the final product after C–Rh bond transformation is considered to be a relatively simple process and could involve C–C/N reductive elimination from the high-valent rhodium complex or protonation of the newly formed C/N–Rh bond.

In order to study the law of chemical reaction rate and find out the intrinsic causes of various chemical reaction rates, synthetic chemists must explore the reaction mechanism and find out the key to determine the reaction rate, so as to control the chemical reaction rate more effectively [104, 106]. Traditional research methods for reaction mechanism include (1) determining the important intermediate or decisive step of a reaction by isotope tracing, (2) determining the effect of different factors (e.g. reaction temperature, solvent, substituent effect, etc.) on reaction rate and selectivity by competitive test, (3) study the relationship between the reaction rate and the concentration of reactants and catalysts obtaining by kinetic experiments, and (4) characterization and tracking of intermediates by instrumental analysis. However, these methods are often macroscopic observation of the average state of many molecules, which cannot watch a process of the transformation for one molecule from a micro perspective. Fortunately, theoretical calculations based on first principles have become one of the important means to study the reaction mechanism with the development of software and the improvement of hardware computing capability in recent several decades. Through theoretical calculation and simulation, the transformation of one molecule in the reaction process can be "watched" more clearly from the microscopic point of view [131, 132]. Actually, theoretical calculation can be considered to be a special kind of microscope, which can see the geometrical structure, electronic structure, spectrum, and dynamic process at atomic level, which is helpful for chemistry to understand the real reaction mechanism.

1.2.2 Mechanistic Study of Rh-Catalyzed C–H Functionalization by Theoretical Methods

Recently, the computational study of Rh-catalyzed C–H functionalization has achieved significant accomplishments [133, 136, 155]. The advances in computational methods and computing power make theoretical calculation a practical and powerful tool for mechanistic study. Both the detailed reaction pathway and origin of selectivity in Rh-catalyzed C–H functionalization can be elucidated through computational study [134, 135, 137]. For instance, Houk and Wu [139] reported density functional theory investigations of the mechanism of the rhodium (III)-catalyzed redox coupling reaction of N-phenoxyacetamides with alkynes. The computational results suggest that the Rh^{III}-Rh^{V}-Rh^{III} mechanism is much more favorable than the Rh^{III}-Rh^{I}-Rh^{III} mechanism. Natural bond orbital analysis confirms the identity of the Rh^{V} intermediate in the catalytic cycle. Chang [107] and co-workers reported a mechanistic study of the rhodium-catalyzed direct C–H amination reaction. Experimental data and DFT calculations reveal that a stepwise pathway involving a key Rh(V)-nitrenoid species that subsequently undergoes amido insertion is favored over a concerted C–N bond formation pathway. DFT calculations and kinetic studies also suggest that the rate-limiting step in the C–H amination reaction is more closely related to the formation of a Rh-nitrene intermediate than the presupposed C–H activation process. Breit [156] and co-workers reported a thorough mechanistic investigation of the rhodium-catalyzed propargylic C–H activation reaction by various spectroscopic and spectrometric methods in combination with DFT calculations. The experimental data and DFT results show that in contrast to the originally proposed mechanism, the catalytic cycle involves intramolecular protonation and not oxidative insertion of rhodium into the O–H bond of the carboxylic acid. In addition, Lan [157] and co-workers reported rhodium/copper-cocatalyzed trans-selective 1,2-diheteroarylation of alkynes with azoles in cooperation with You. The calculated results show that the catalytic cycle involves C–H bond activation, alkyne insertion, transmetallation with aryl-Cu, formation of a second C–C bond via an unpredictable trans-nucleophilic addition, and two single electron transfer steps.

Although the tremendous progress in theoretical studies has increased the understanding of the mechanism of rhodium-catalyzed C–H bond functionalization, these studies are often case by case discussions. There have been few reviews of the computational advances in this area, which leads chemists to not having a systematic knowledge of the mechanism. Therefore, there is an urgent need for a new review detailing the recent computational progress in rhodium-catalyzed C–H bond functionalization. Here, we choose to focus on the theoretical aspects of rhodium-catalyzed C–H bond functionalization. Our review will provide the first comprehensive and systematical summary of the theoretical advances in rhodium-catalyzed C–H functionalization in the past decade. In this context, arylation, alkylation, vinylation, alkynylation, carbonylation, hydroacylation, and cyclization catalyzed by Rh are discussed. In each part, the order of the elementary reactions, change in the oxidation state of Rh, and its relationship with the C–H bond activation mode are emphasized to reveal the mechanistic diversity.

References

1. Arndtsen BA, Bergman RG, Mobley TA, Peterson TH (2002) Selective intermolecular carbon-hydrogen bond activation by synthetic metal complexes in homogeneous solution. Acc Chem Res 28:154–162
2. Cheng D, Ishihara Y, Tan B, Barbas CF (2014) Organocatalytic asymmetric assembly reactions: synthesis of spirooxindoles via organocascade strategies. ACS Catal 4:743–762
3. Crisenza GEM, Mazzarella D, Melchiorre P (2020) Synthetic methods driven by the photoactivity of electron donor-acceptor complexes. J Am Chem Soc 142:5461–5476
4. Kendall JL, Canelas DA, Young JL, DeSimone JM (1999) Polymerizations in supercritical carbon dioxide. Chem Rev 99:543–564
5. Kim J, Kim H, Park SB (2014) Privileged structures: efficient chemical "navigators" toward unexplored biologically relevant chemical spaces. J Am Chem Soc 136:14629–14638
6. Melchiorre P, Marigo M, Carlone A, Bartoli G (2008) Asymmetric aminocatalysis-gold rush in organic chemistry. Angew Chem Int Ed 47:6138–6171
7. Nicolaou KC, Hale CRH, Nilewski C, Ioannidou HA (2012) Constructing molecular complexity and diversity: total synthesis of natural products of biological and medicinal importance. Chem Soc Rev 41:5185
8. Santaniello E, Ferraboschi P, Grisenti P, Manzocchi A (1992) The biocatalytic approach to the preparation of enantiomerically pure chiral building blocks. Chem Rev 92:1071–1140
9. Segura JL, Martin N (1999) o-Quinodimethanes: efficient intermediates in organic synthesis. Chem Rev 99:3199–3246
10. Tortajada A, Julia-Hernandez F, Borjesson M, Moragas T, Martin R (2018) Transition-metal-catalyzed carboxylation reactions with carbon dioxide. Angew Chem Int Ed 57:15948–15982
11. Whitesides GM, Simanek EE, Mathias JP, Seto CT, Chin D, Mammen M, Gordon DM (2002) Noncovalent synthesis: using physical-organic chemistry to make aggregates. Acc Chem Res 28:37–44
12. Zhan G, Du W, Chen Y-C (2017) Switchable divergent asymmetric synthesis via organocatalysis. Chem Soc Rev 46:1675–1692
13. Zimmer R, Dinesh CU, Nandanan E, Khan FA (2000) Palladium-catalyzed reactions of allenes. Chem Rev 100:3067–3126
14. Liu S, Lei Y, Li Y, Zhang T, Wang H, Lan Y (2014) Hexahapto-chromium complexes of graphene: a theoretical study. RSC Adv. 4:28640–28644
15. Yudin AK (2010) Catalyzed carbon-heteroatom bond formation. Wiley-VCH, Weinheim
16. Katritzky A, Rees CW, Scriven EF, Ramsden C, Taylor R (1984) Comprehensive heterocyclic chemistry. Elsevier Science, Oxford
17. Monnier F, Taillefer M (2009) Catalytic C–C, C–N, and C–O Ullmann-type coupling reactions. Angew Chem Int Ed 48:6954–6971
18. Nasrollahzadeh M, Issaabadi Z, Tohidi MM, Mohammad Sajadi S (2018) Recent progress in application of graphene supported metal nanoparticles in C–C and C–X coupling reactions. Chem Rec 18:165–229
19. Sherry BD, Furstner A (2008) The promise and challenge of iron-catalyzed cross coupling. Acc Chem Res 41:1500–1511
20. Tang X, Wu W, Zeng W, Jiang H (2018) Copper-catalyzed oxidative carbon-carbon and/or carbon-heteroatom bond formation with O_2 or internal oxidants. Acc Chem Res 51:1092–1105
21. Valente C, Calimsiz S, Hoi KH, Mallik D, Sayah M, Organ MG (2012) The development of bulky palladium NHC complexes for the most-challenging cross-coupling reactions. Angew Chem Int Ed 51:3314–3332

22. Wolfe JP, Wagaw S, Marcoux JF, Buchwald SL (1998) Rational development of practical catalysts for aromatic carbon-nitrogen bond formation. Acc Chem Res 31:805–818
23. Wu W, Jiang H (2012) Palladium-catalyzed oxidation of unsaturated hydrocarbons using molecular oxygen. Acc Chem Res 45:1736–1748
24. Xiao Q, Zhang Y, Wang J (2013) Diazo compounds and N-tosylhydrazones: novel cross-coupling partners in transition-metal-catalyzed reactions. Acc Chem Res 46:236–247
25. Yu D-G, Li B-J, Shi Z-J (2010) Exploration of new C–O electrophiles in cross-coupling reactions. Acc Chem Res 43:1486–1495
26. Liu C, Ji CL, Zhou T, Hong X, Szostak M (2019) Decarbonylative phosphorylation of carboxylic acids via redox-neutral palladium catalysis. Org Lett 21:9256–9261
27. Mo Y, Lu Z, Rughoobur G, Patil P, Gershenfeld N, Akinwande AI, Buchwald SL, Jensen KF (2020) Microfluidic electrochemistry for single-electron transfer redox-neutral reactions. Science 368:1352–1357
28. Nielsen MK, Shields BJ, Liu J, Williams MJ, Zacuto MJ, Doyle AG (2017) Mild, redox-neutral formylation of aryl chlorides through the photocatalytic generation of chlorine radicals. Angew Chem Int Ed 56:7191–7194
29. Qi L-W, Li S, Xiang S-H, Wang J, Tan B (2019) Asymmetric construction of atropisomeric biaryls via a redox neutral cross-coupling strategy. Nat Catal 2:314–323
30. Xie J, Jin H, Hashmi ASK (2017) The recent achievements of redox-neutral radical C–C cross-coupling enabled by visible-light. Chem Soc Rev 46:5193–5203
31. Xuan J, Zeng TT, Feng ZJ, Deng QH, Chen JR, Lu LQ, Xiao WJ, Alper H (2015) Redox-neutral alpha-allylation of amines by combining palladium catalysis and visible-light photoredox catalysis. Angew Chem Int Ed 54:1625–1628
32. Zhou C, Lei T, Wei X-Z, Ye C, Liu Z, Chen B, Tung CH, Wu L-Z (2020) Metal-Free, redox-neutral, site-selective access to heteroarylamine via direct radical-radical cross-coupling powered by visible light photocatalysis. J Am Chem Soc 142:16805–16813
33. Zhu C, Yue H, Nikolaienko P, Rueping M (2020) Merging electrolysis and nickel catalysis in redox neutral cross-coupling reactions: experiment and computation for electrochemically induced C–P and C–Se bonds formation. CCS Chemistry 2:179–190
34. Guan H, Zhang Q, Walsh PJ, Mao J (2020) Nickel/Photoredox-catalyzed asymmetric reductive cross-coupling of racemic alpha-chloro esters with aryl iodides. Angew Chem Int Ed 59:5172–5177
35. Kadunce NT, Reisman SE (2015) Nickel-catalyzed asymmetric reductive cross-coupling between heteroaryl iodides and α-Chloronitriles. J Am Chem Soc 137:10480–10483
36. Li Y, Luo Y, Peng L, Li Y, Zhao B, Wang W, Pang H, Deng Y, Bai R, Lan Y, Yin G (2020) Reaction scope and mechanistic insights of nickel-catalyzed migratory Suzuki-Miyaura cross-coupling. Nat Commun 11:417
37. Mitsui A, Nagao K, Ohmiya H (2020) Copper-Catalyzed enantioselective reductive cross-coupling of aldehydes and imines. Org Lett 22:800–803
38. Poremba KE, Dibrell SE, Reisman SE (2020) Nickel-Catalyzed enantioselective reductive cross-coupling reactions. ACS Catal 10:8237–8246
39. Tollefson EJ, Erickson LW, Jarvo ER (2015) Stereospecific intramolecular reductive cross-electrophile coupling reactions for cyclopropane synthesis. J Am Chem Soc 137:9760–9763
40. Zhong YW, Dong YZ, Fang K, Izumi K, Xu MH, Lin G-Q (2005) A highly efficient and direct approach for synthesis of enantiopure beta-amino alcohols by reductive cross-coupling of chiral N-tert-butanesulfinyl imines with aldehydes. J Am Chem Soc 127:11956–11957
41. Ashenhurst JA (2010) Intermolecular oxidative cross-coupling of arenes. Chem Soc Rev 39:540–548
42. Chen M, Zheng X, Li W, He J, Lei A (2010) Palladium-catalyzed aerobic oxidative cross-coupling reactions of terminal alkynes with alkylzinc reagents. J Am Chem Soc 132:4101–4103
43. Hull KL, Sanford MS (2009) Mechanism of benzoquinone-promoted palladium-catalyzed oxidative cross-coupling reactions. J Am Chem Soc 131:9651–9653

44. Liu C, Liu D, Lei A (2014) Recent advances of transition-metal catalyzed radical oxidative cross-couplings. Acc Chem Res 47:3459–3470
45. Liu C, Zhang H, Shi W, Lei A (2011) Bond formations between two nucleophiles: transition metal catalyzed oxidative cross-coupling reactions. Chem Rev 111:1780–1824
46. Shi W, Liu C, Lei A (2011) Transition-metal catalyzed oxidative cross-coupling reactions to form C–C bonds involving organometallic reagents as nucleophiles. Chem Soc Rev 40:2761–2776
47. Yuan Y, Lei A (2019) Electrochemical oxidative cross-coupling with hydrogen evolution reactions. Acc Chem Res 52:3309–3324
48. McNaught AD, Wilkinson A (1997) Compendium of chemical terminology: IUPAC recommendations. Blackwell Science, Oxford, Malden
49. Cartmell E, Fowles GWA (1996) Valency and molecular structure. D Van Nostrand Co, Princeton
50. Scudder PH (2013) Electron flow in organic chemistry: a decision based guide to organic mechanisms. Wiley, New York
51. Cho YH, Zunic V, Senboku H, Olsen M, Lautens M (2006) Rhodium-catalyzed ring-opening reactions of N-boc-azabenzonorbornadienes with amine nucleophiles. J Am Chem Soc 128:6837–6846
52. Hartwig JF (2008) Evolution of a fourth generation catalyst for the amination and thioetherification of aryl halides. Acc Chem Res 41:1534–1544
53. Hartwig JF (2008) Carbon-heteroatom bond formation catalysed by organometallic complexes. Nature 455:314–322
54. Staubitz A, Robertson AP, Sloan ME, Manners I (2010) Amine- and phosphine-borane adducts: new interest in old molecules. Chem Rev 110:4023–4078
55. Watanabe Y, Morisaki Y, Kondo T, Mitsudo Ta TA (1996) Ruthenium complex-controlled catalytic N-Mono- or N, N-Dialkylation of heteroaromatic amines with alcohols. J Org Chem 61:4214–4218
56. Domingo LR, Saez JA, Zaragoza RJ, Arno M (2008) Understanding the participation of quadricyclane as nucleophile in polar [2sigma + 2sigma + 2pi] cycloadditions toward electrophilic pi molecules. J Org Chem 73:8791–8799
57. Muller TE, Hultzsch KC, Yus M, Foubelo F, Tada M (2008) Hydroamination: direct addition of amines to alkenes and alkynes. Chem Rev 108:3795–3892
58. Rosenberg L (2013) Mechanisms of metal-catalyzed hydrophosphination of alkenes and alkynes. ACS Catal 3:2845–2855
59. Canty AJ, van Koten G (2002) Mechanisms of d8 organometallic reactions involving electrophiles and intramolecular assistance by nucleophiles. Acc Chem Res 28:406–413
60. Gring M, Kuhnert M, Langen T, Kitagawa T, Rauer B, Schreitl M, Mazets I, Smith DA, Demler E, Schmiedmayer J (2012) Relaxation and prethermalization in an isolated quantum system. Science 337:1318–1322
61. Gunay A, Theopold KH (2010) C–H bond activations by metal oxo compounds. Chem Rev 110:1060–1081
62. Hashiguchi BG, Bischof SM, Konnick MM, Periana RA (2012) Designing catalysts for functionalization of unactivated C–H bonds based on the CH activation reaction. Acc Chem Res 45:885–898
63. Hashiguchi BG, Young KJ, Yousufuddin M, Goddard WA 3rd, Periana RA (2010) Acceleration of nucleophilic CH activation by strongly basic solvents. J Am Chem Soc 132:12542–12545
64. Liu C, Yuan J, Gao M, Tang S, Li W, Shi R, Lei A (2015) Oxidative coupling between two hydrocarbons: an update of recent C–H functionalizations. Chem Rev 115:12138–12204
65. Zhu D, Chen L, Fan H, Yao Q, Zhu S (2020) Recent progress on donor and donor-donor carbenes. Chem Soc Rev 49:908–950
66. Cheng W-M, Shang R (2020) Transition metal-catalyzed organic reactions under visible light: recent developments and future perspectives. ACS Catal 10:9170–9196

67. Kaga A, Chiba S (2017) Engaging radicals in transition metal-catalyzed cross-coupling with alkyl electrophiles: recent advances. ACS Catal 7:4697–4706
68. Lautens M, Klute W, Tam W (1996) Transition metal-mediated cycloaddition reactions. Chem Rev 96:49–92
69. Müller TE, Grosche M, Herdtweck E, Pleier A-K, Walter E, Yan Y-K (2000) Developing transition-metal catalysts for the intramolecular hydroamination of alkynes. Organometallics 19:170–183
70. Park Y, Kim Y, Chang S (2017) Transition metal-catalyzed C–H amination: scope, mechanism, and applications. Chem Rev 117:9247–9301
71. Park YJ, Park JW, Jun CH (2008) Metal-organic cooperative catalysis in C–H and C–C bond activation and its concurrent recovery. Acc Chem Res 41:222–234
72. Saint-Denis TG, Zhu RY, Chen G, Wu QF, Yu JQ (2018) Enantioselective C(sp(3))H bond activation by chiral transition metal catalysts. Science 359
73. Wang F, Yu S, Li X (2016) Transition metal-catalysed couplings between arenes and strained or reactive rings: combination of C–H activation and ring scission. Chem Soc Rev 45:6462–6477
74. Xu X, Doyle MP (2014) The [3 + 3]-cycloaddition alternative for heterocycle syntheses: catalytically generated metalloenolcarbenes as dipolar adducts. Acc Chem Res 47:1396–1405
75. Kakiuchi F, Murai S, Murai S, Alper H, Gossage RA, Grushin VV, Hidai M, Ito Y, Jones WD, Kakiuchi F, van Koten G, Lin YS (1999) Activation of C–H bonds: catalytic reactions. Activation of unreactive bonds and organic synthesis. Springer, Berlin, pp 47–79
76. Huang Z, Lim HN, Mo F, Young MC, Dong G (2015) Transition metal-catalyzed ketone-directed or mediated C–H functionalization. Chem Soc Rev 44:7764–7786
77. Ritleng V, Sirlin C, Pfeffer M (2002) Ru-, Rh-, and Pd-catalyzed C–C bond formation involving C–H activation and addition on unsaturated substrates: reactions and mechanistic aspects. Chem Rev 102:1731–1770
78. Shilov AE, Shul'pin GB (1997) Activation of cminus signH bonds by metal complexes. Chem Rev 97:2879–2932
79. Alberico D, Scott ME, Lautens M (2007) Aryl-aryl bond formation by transition-metal-catalyzed direct arylation. Chem Rev 107:174–238
80. Bellina F, Rossi R (2010) Transition metal-catalyzed direct arylation of substrates with activated $sp3$-hybridized C–H bonds and some of their synthetic equivalents with aryl halides and pseudohalides. Chem Rev 110:1082–1146
81. Chen F, Wang T, Jiao N (2014) Recent advances in transition-metal-catalyzed functionalization of unstrained carbon-carbon bonds. Chem Rev 114:8613–8661
82. Cho SH, Kim JY, Kwak J, Chang S (2011) Recent advances in the transition metal-catalyzed twofold oxidative C–H bond activation strategy for C–C and C–N bond formation. Chem Soc Rev 40:5068–5083
83. Gao DW, Gu Q, Zheng C, You SL (2017) Synthesis of planar chiral ferrocenes via transition-metal-catalyzed direct C–H bond functionalization. Acc Chem Res 50:351–365
84. Iwai T, Sawamura M (2015) Transition-metal-catalyzed site-selective C–H functionalization of quinolines beyond C2 selectivity. ACS Catal 5:5031–5040
85. Miao J, Ge H (2015) Recent advances in first-row-transition-metal-catalyzed Dehydrogenative cou-pling of C($sp3$)-H Bonds. European J Org Chem 2015:7859–7868
86. Zhang X-S, Chen K, Shi Z-J (2014) Transition metal-catalyzed direct nucleophilic addition of C–H bonds to carbon–heteroatom double bonds. Chem Sci 5:2146–2159
87. He J, Wasa M, Chan KSL, Shao Q, Yu JQ (2017) Palladium-catalyzed transformations of alkyl C–H bonds. Chem Rev 117:8754–8786
88. Lewis JC, Bergman RG, Ellman JA (2008) Direct functionalization of nitrogen heterocycles via Rh-catalyzed C–H bond activation. Acc Chem Res 41:1013–1025
89. Newton CG, Wang SG, Oliveira CC, Cramer N (2017) Catalytic enantioselective transformations involving C–H bond cleavage by transition-metal complexes. Chem Rev 117:8908–8976
90. Shan C, Zhu L, Qu LB, Bai R, Lan Y (2018) Mechanistic view of Ru-catalyzed C–H bond activation and functionalization: computational advances. Chem Soc Rev 47:7552–7576

91. Su B, Cao ZC, Shi ZJ (2015) Exploration of earth-abundant transition metals (Fe Co, and Ni) as catalysts in unreactive chemical bond activations. Acc Chem Res 48:886–896
92. Wang F, Chen P, Liu G (2018) Copper-catalyzed radical relay for asymmetric radical transformations. Acc Chem Res 51:2036–2046
93. Colby DA, Bergman RG, Ellman JA (2010) Rhodium-catalyzed C–C bond formation via heteroatom-directed C–H bond activation. Chem Rev 110:624–655
94. Colby DA, Tsai AS, Bergman RG, Ellman JA (2012) Rhodium catalyzed chelation-assisted C–H bond functionalization reactions. Acc Chem Res 45:814–825
95. Berman AM, Lewis JC, Bergman RG, Ellman JA (2008) Rh(I)-catalyzed direct arylation of pyridines and quinolines. J Am Chem Soc 130:14926–14927
96. Colby DA, Bergman RG, Ellman JA (2008) Synthesis of dihydropyridines and pyridines from imines and alkynes via C–H activation. J Am Chem Soc 130:3645–3651
97. Hesp KD, Bergman RG, Ellman JA (2011) Expedient synthesis of N-acyl anthranilamides and beta-enamine amides by the Rh(III)-catalyzed amidation of aryl and vinyl C–H bonds with isocyanates. J Am Chem Soc 133:11430–11433
98. Lewis JC, Bergman RG, Ellman JA (2007) Rh(I)-catalyzed alkylation of quinolines and pyridines via C–H bond activation. J Am Chem Soc 129:5332–5333
99. Lewis JC, Berman AM, Bergman RG, Ellman JA (2008) Rh(I)-catalyzed arylation of heterocycles via C–H bond activation: expanded scope through mechanistic insight. J Am Chem Soc 130:2493–2500
100. Lian Y, Bergman RG, Lavis LD, Ellman JA (2013) Rhodium(III)-catalyzed indazole synthesis by C–H bond functionalization and cyclative capture. J Am Chem Soc 135:7122–7125
101. Tsai AS, Tauchert ME, Bergman RG, Ellman JA (2011) Rhodium(III)-catalyzed arylation of Boc-imines via C–H bond functionalization. J Am Chem Soc 133:1248–1250
102. Wilson RM, Thalji RK, Bergman RG, Ellman JA (2006) Enantioselective synthesis of a PKC inhibitor via catalytic C–H bond activation. Org Lett 8:1745–1747
103. O'Malley SJ, Tan KL, Watzke A, Bergman RG, Ellman JA (2005) Total synthesis of (+)-lithospermic acid by asymmetric intramolecular alkylation via catalytic C–H bond activation. J Am Chem Soc 127:13496–13497
104. Schroder N, Lied F, Glorius F (2015) Dual role of Rh(III) catalyst enables regioselective halogenation of (electron-rich) heterocycles. J Am Chem Soc 137:1448–1451
105. Song G, Li X (2015) Substrate activation strategies in rhodium(III)-catalyzed selective functionalization of arenes. Acc Chem Res 48:1007–1020
106. Zhang X, Qi Z, Li X (2014) Rhodium(III)-catalyzed C–C and C–O coupling of quinoline N-oxides with alkynes: combination of C–H activation with O-atom transfer. Angew Chem Int Ed 53:10794–10798
107. Shin K, Kim H, Chang S (2015) Transition-metal-catalyzed C-N bond forming reactions using organic azides as the nitrogen source: a journey for the mild and versatile C–H amination. Acc Chem Res 48:1040–1052
108. Ye B, Cramer N (2012) Chiral cyclopentadienyl ligands as stereocontrolling element in asymmetric C–H functionalization. Science 338:504–506
109. Hyster TK, Rovis T (2010) Rhodium-catalyzed oxidative cycloaddition of benzamides and alkynes via C–H/N–H activation. J Am Chem Soc 132:10565–10569
110. Hyster TK, Ruhl KE, Rovis T (2013) A coupling of benzamides and donor/acceptor diazo compounds to form gamma-lactams via Rh(III)-catalyzed C–H activation. J Am Chem Soc 135:5364–5367
111. Piou T, Rovis T (2018) Electronic and steric tuning of a prototypical piano stool complex: Rh(III) catalysis for C–H functionalization. Acc Chem Res 51:170–180
112. Ackermann L (2011) Carboxylate-assisted transition-metal-catalyzed C–H bond functionalizations: mechanism and scope. Chem Rev 111:1315–1345
113. Ackermann L, Vicente R, Kapdi AR (2009) Transition-metal-catalyzed direct arylation of (hetero)arenes by C–H bond cleavage. Angew Chem Int Ed 48:9792–9826
114. Dyker G (1999) Transition metal catalyzed coupling reactions under C–H activation. Angew Chem Int Ed 38:1698–1712

115. Giri R, Shi BF, Engle KM, Maugel N, Yu JQ (2009) Transition metal-catalyzed C–H activation reactions: diastereoselectivity and enantioselectivity. Chem Soc Rev 38:3242–3272

116. Trost BM, Van Vranken DL (1996) Asymmetric transition metal-catalyzed allylic alkylations. Chem Rev 96:395–422

117. Lim YG, Kim YH, Kang JB (1994) Rhodium-catalysed regioselective alkylation of the phenyl ring of 2-phenylpyridines with olefins. J Chem Soc Chem Commun:2267

118. Jun C-H, Hong J-B, Kim Y-H, Chung K-Y (2000) The Catalytic alkylation of aromatic imines by Wilkinson's complex: the domino reaction of hydroacylation andortho-Alkylation. Angew Chem Int Ed 39:3440–3442

119. Lim Y-G, Han J-S, Koo BT, Kang J-B (2004) Regioselective alkylation of aromatic aldimines and ketimines via C–H bond activation by a rhodium catalyst. J Mol Catal A-Chem 209:41–49

120. Lim Y-G, Lee K-H, Koo BT, Kang J-B (2001) Rhodium(I)-catalyzed ortho- alkenylation of 2-phenylpyridines with alkynes. Tetrahedron Lett 42:7609–7612

121. Lim SG, Lee JH, Moon CW, Hong J-B, Jun CH (2003) Rh(I)-catalyzed direct ortho-alkenylation of aromatic ketimines with alkynes and its application to the synthesis of isoquinoline derivatives. Org Lett 5:2759–2761

122. Oi S, Fukita S, Inoue Y (1998) Rhodium-catalysed direct ortho arylation of 2-arylpyridines with arylstannanes via C–H activation. Chem Commun:2439–2440

123. Asaumi T, Matsuo T, Fukuyama T, Ie Y, Kakiuchi F, Chatani N (2004) Ruthenium- and rhodium-catalyzed direct carbonylation of the ortho C-H bond in the benzene ring of N-arylpyrazoles. J Org Chem 69:4433–4440

124. Jun C-H, Moon CW, Kim Y-M, Lee H, Lee JH (2002) Chelation-assisted β-alkylation of α, β-unsaturated ketone using Rh(I) catalyst and dialkyl amine. Tetrahedron Lett 43:4233–4236

125. Aissa C, Furstner A (2007) A rhodium-catalyzed C–H activation/cycloisomerization tandem. J Am Chem Soc 129:14836–14837

126. Tan KL, Bergman RG, Ellman JA (2001) Annulation of alkenyl-substituted heterocycles via rhodium-catalyzed intramolecular C–H activated coupling reactions. J Am Chem Soc 123:2685–2686

127. Lewis JC, Wiedemann SH, Bergman RG, Ellman JA (2004) Arylation of heterocycles via rhodium-catalyzed C–H bond functionalization. Org Lett 6:35–38

128. Ueura K, Satoh T, Miura M (2007) Rhodium- and iridium-catalyzed oxidative coupling of benzoic acids with alkynes via regioselective C–H bond cleavage. J Org Chem 72:5362–5367

129. Chen W-J, Lin Z (2014) Rhodium(III)-catalyzed hydrazine-directed C–H activation for indole synthesis: mechanism and role of internal oxidant probed by DFT studies. Organometallics 34:309–318

130. Sperger T, Sanhueza IA, Kalvet I, Schoenebeck F (2015) Computational Studies of synthetically relevant homogeneous organometallic catalysis involving Ni, Pd, Ir, and Rh: an overview of commonly employed DFT methods and mechanistic insights. Chem Rev 115:9532–9586

131. Lin Z (2010) Interplay between theory and experiment: computational organometallic and transition metal chemistry. Acc Chem Res 43:602–611

132. Xue L, Lin Z (2010) Theoretical aspects of palladium-catalysed carbon-carbon cross-coupling reactions. Chem Soc Rev 39:1692–1705

133. Dang Y, Qu S, Tao Y, Deng X, Wang ZX (2015) Mechanistic insight into ketone alpha-alkylation with unactivated olefins via C–H activation promoted by metal-organic cooperative catalysis (MOCC): enriching the MOCC chemistry. J Am Chem Soc 137:6279–6291

134. Guimond N, Gorelsky SI, Fagnou K (2011) Rhodium(III)-catalyzed heterocycle synthesis using an internal oxidant: improved reactivity and mechanistic studies. J Am Chem Soc 133:6449–6457

135. Li Y, Liu S, Qi Z, Qi X, Li X, Lan Y (2015) The mechanism of N–O bond cleavage in Rhodium-Catalyzed C–H bond functionalization of quinoline N-oxides with alkynes: a computational study. Chem A Eur J 21:10131–10137

136. Neufeldt SR, Jimenez-Oses G, Huckins JR, Thiel OR, Houk KN (2015) Pyridine N-Oxide vs pyridine substrates for Rh(III)-Catalyzed oxidative C–H Bond Functionalization. J Am Chem Soc 137:9843–9854

137. Wodrich MD, Ye B, Gonthier JF, Corminboeuf C, Cramer N (2014) Ligand-controlled regiodivergent pathways of rhodium(III)-catalyzed dihydroisoquinolone synthesis: experimental and computational studies of different cyclopentadienyl ligands. Chem A Eur J 20:15409–15418

138. Wu JQ, Zhang SS, Gao H, Qi Z, Zhou CJ, Ji WW, Liu Y, Chen Y, Li Q, Li X, Wang H (2017) Experimental and theoretical studies on rhodium-catalyzed coupling of benzamides with 2,2-Difluorovinyl tosylate: diverse synthesis of fluorinated heterocycles. J Am Chem Soc 139:3537–3545

139. Yang YF, Houk KN, Wu YD (2016) Computational exploration of Rh(III)/Rh(V) and Rh(III)/Rh(I) catalysis in Rhodium(III)-Catalyzed C–H activation reactions of N-Phenoxyacetamides with alkynes. J Am Chem Soc 138:6861–6868

140. Yu S, Liu S, Lan Y, Wan B, Li X (2015) Rhodium-catalyzed C–H activation of phenacyl ammonium salts assisted by an oxidizing C–N bond: a combination of experimental and theoretical studies. J Am Chem Soc 137:1623–1631

141. Garcia-Cuadrado D, Braga AA, Maseras F, Echavarren AM (2006) Proton abstraction mechanism for the palladium-catalyzed intramolecular arylation. J Am Chem Soc 128:1066–1067

142. Lafrance M, Rowley CN, Woo TK, Fagnou K (2006) Catalytic intermolecular direct arylation of perfluorobenzenes. J Am Chem Soc 128:8754–8756

143. Hoyano JK, Graham WAG (1982) Oxidative addition of the carbon-hydrogen bonds of neopentane and cyclohexane to a photochemically generated iridium(I) complex. J Am Chem Soc 104:3723–3725

144. Pivsa-Art S, Satoh T, Kawamura Y, Miura M, Nomura M (1998) Palladium-catalyzed arylation of azole compounds with aryl halides in the presence of alkali metal carbonates and the use of copper iodide in the reaction. Bull Chem Soc Japan 71:467–473

145. Watson PL (1983) Methane exchange reactions of lanthanide and early-transition-metal methyl complexes. J Am Chem Soc 105:6491–6493

146. Guan ZH, Ren ZH, Spinella SM, Yu S, Liang YM, Zhang X (2009) Rhodium-catalyzed direct oxidative carbonylation of aromatic C–H bond with CO and alcohols. J Am Chem Soc 131:729–733

147. Ishii Y, Chatani N, Kakiuchi F, Murai S (1997) Rhodium-catalyzed reaction of N-(2-Pyridinyl)piperazines with CO and ethylene. a novel carbonylation at a C − H bond in the piperazine ring. Organometallics 16:3615–3622

148. Deng H, Li H, Wang L (2015) A unique alkylation of azobenzenes with allyl acetates by Rh(III)-Catalyzed C–H functionalization. Org Lett 17:2450–2453

149. Huang L, Wang Q, Qi J, Wu X, Huang K, Jiang H (2013) Rh(iii)-catalyzed ortho-oxidative alkylation of unactivated arenes with allylic alcohols. Chem Sci 4:2665

150. Shibata K, Chatani N (2014) Rhodium-catalyzed alkylation of C-H bonds in aromatic amides with alpha, beta-unsaturated esters. Org Lett 16:5148–5151

151. Dateer RB, Chang S (2015) Selective cyclization of arylnitrones to indolines under external oxidant-free conditions: dual role of Rh(III) catalyst in the C–H activation and oxygen atom transfer. J Am Chem Soc 137:4908–4911

152. Sharma U, Park Y, Chang S (2014) Rh(III)-catalyzed traceless coupling of quinoline N-oxides with internal diarylalkynes. J Org Chem 79:9899–9906

153. Chen X, Zheng G, Li Y, Song G, Li X (2017) Rhodium-catalyzed site-selective coupling of indoles with diazo esters: C4-alkylation versus C2-Annulation. Org Lett 19:6184–6187

154. Wu Y, Chen Z, Yang Y, Zhu W, Zhou B (2018) Rh(III)-catalyzed redox-neutral unsymmetrical c–h alkylation and amidation reactions of N-Phenoxyacetamides. J Am Chem Soc 140:42–45

155. Qi X, Li Y, Bai R, Lan Y (2017) Mechanism of rhodium-catalyzed C–H functionalization: advances in theoretical investigation. Acc Chem Res 50:2799–2808

156. Vautravers NR, Regent DD, Breit B (2011) Inter- and intramolecular hydroacylation of alkenes employing a bifunctional catalyst system. Chem Commun 47:6635–6637

157. Tan G, Zhu L, Liao X, Lan Y, You J (2017) Rhodium/copper cocatalyzed highly trans-selective 1,2-diheteroarylation of alkynes with azoles via c-h addition/oxidative cross-coupling: a combined experimental and theoretical study. J Am Chem Soc 139:15724–15737

Chapter 2
Computational Methods in Rh-Catalyzed C–H Functionalization

Cheng-Xing Cui, Song Liu, Chun-Xiang Li, Ruopeng Bai, and Yu Lan

2.1 DFT Method

The theoretical treatment of a polyatomic molecular involves the ab initio method, the semiempirical method, the density-functional theory (DFT) method, and the molecular-mechanics method. An *ab initio* method [1–4] calculates the electronic wavefunction of a molecule by using the correct Hamiltonian [5] to solve the Schrödinger equation. However, a semiempirical molecular quantum mechanical method [6, 7] uses a simpler Hamiltonian than the correct molecular Hamiltonian. In the semiempirical method, experimental data or results from ab initio calculations were adopted to adjust the values of parameters. A DFT method [8, 9] calculates the molecular electron probability density ρ rather than molecular wavefunction and then calculates the molecular electronic energy from ρ. A molecular mechanics method [10] views the polyatomic molecule as a collection of atoms, which are held together by bonds. The molecular mechanics method obtains the molecular energy without wavefunction but in terms of force constants for bond bending, stretching, torsion, and other parameters. Consequently, the molecular mechanics method is not a quantum mechanical method.

The solution of Schrödinger equation with ab initio method needs to deal with $3n$ variables for an n-electron system. Meanwhile, only three variables should be considered for the electron probability density ρ in the DFT method. As a result, for the computational studies of Rh-catalyzed C–H bond activation, DFT method was always the first choice, which could give an adequate compromise between computational consumption and accuracy.

There is no unified standard for the classification of density functionals in the physical chemistry field. As shown in Fig. 2.1, J. P. Perdew proposed using the "Jacob's ladder" to classify the level of functionals [11]. The ground in "Jacob's ladder" is HF theory, which is an imprecise method including neither exchange energy nor correlation energy. The first rung in "Jacob's ladder" is the functional based on L(S)DA, the variable in this kind of functional is the local spin density. The second rung in

© The Author(s), under exclusive license to Springer Nature Singapore Pte Ltd. 2021
Y. Lan et al., *Computational Advances of Rh-Catalyzed C–H Functionalization*,
SpringerBriefs in Molecular Science,
https://doi.org/10.1007/978-981-16-0432-4_2

Fig. 2.1 The Jacob's Ladder of density functionals

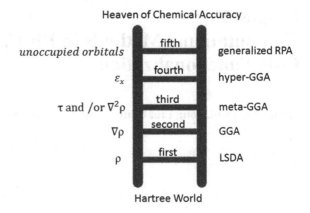

"Jacob's ladder" is the GGA. The variables in this kind of functionals are local spin density and its gradient. There are no analytic expressions for both exchange functionals and correlation functionals of GGA density functionals. The third rung in "Jacob's ladder" of density functionals is meta-GGA functionals. The variables with more functionals than GGA are the kinetic energy density or the second derivative of the local spin density. The most common meta-GGA involved M06-L [12], TPSS [13], and VSXC [14], which are often used in computational organometallic chemistry currently. The fourth rung is hybrid-GGA and hybrid-meta GGA. This kind of functionals are the most popular functional in computational chemistry currently, into which HF exchange is introduced. In the field of computational organometallic chemistry, the commonly used hybrid-GGA functionals involve B3-LYP [15, 16], B97 [17, 18], O3LYP [19], PBE0 [20, 21], mPW1PW [22], and X3LYP [23]; the commonly used hybrid-meta GGA functionals involve M05 [24], M05-2X [25], M06 [26], M06-HF [27], M06-2X [26], TPSSh [13], and MPW1K [28, 29]. In the fifth rung of "Jacob's ladder", the information of virtual orbital is used to build double-hybrid functional. For example, in this type of functionals, B2PLYP [30] and mPW2PLYP [31], Kohn–Sham unoccupied orbitals are used to calculate MP2-type correlation functional [32].

Both geometry and energy information are important for understanding the mechanisms of Rh-catalyzed C–H bond activation [33]. To obtain reliable thermodynamic information from theoretical calculations, accurate molecular geometry is especially important, which is the basis for the energy and other property calculations [34]. The most popular functional may be one of the fourth rung functionals, i.e. B3-LYP, which combines the standard local exchange functional with the gradient correction of Becke [35] and uses the Lee–Yang–Parr [36] correlation functional and is always the preferred functional for the geometry optimization for the theoretical investigation of Rh-catalyzed C–H bond activation [37, 38]. As been mentioned above, there are hundreds of density functionals, many of which would perform better than B3-LYP in their own field of expertise [34, 39]. However, very few of them have more comprehensive performance than B3-LYP, which leads to the popularity of that functional. B3-LYP functional has two main weaknesses, including disregard of dispersion interaction and the bad performance on charge transfer and Rydberg excitations [40]. The first problem can be completed modified by DFT-D3 correction without additional computing time [41, 42]. The second one can be solved by the variant CAM-B3-LYP functional [43]. These amendments continue to extend the service life of B3-LYP functional. Although B3-LYP performs well in geometry optimizations, it is not suitable for energy calculations in Rh-catalyzed C–H functionalization.

From many calculation results, M06, which is a Minnesota series functional proposed by Truhlar in 2007, is one of the best alternative functionals of B3-LYP [12, 26]. The M06 is not only suitable for geometry optimizations but also well behaved in energy calculations in Rh-catalyzed C–H functionalization [44, 45]. In this functional, dispersion effect is introduced in its fitting parameters, which reveals good weak interactions [26]. The 54% HF exchange component leads to better performance on the calculation of charge transfer and Rydberg excitation. The shortcoming is that Minnesota series functionals require much higher accuracy of DFT integration grid than B3-LYP [46]. It could be solved by the improvement of that, however, it will obviously increase the time-consuming. Moreover, M06 functional is parameterized for the main group elements, therefore, it might be unsuited for the calculation of transition metal involved system. As an alternative, M06-L [12] and M11-L [47] functional can be used in this case. It is worth noting that M06-L and M11-L are designed to capture the main dependence of the exchange-correlation energy on local spin density, spin density gradient, and spin kinetic energy density. Moreover, M06-L and M11-L are parameterized to satisfy the uniform electron gas limit. Therefore, M06-L and M11-L have good performance for energy calculations in Rh-catalyzed C–H functionalization reaction [48].

The ωB97XD functional [49] is another alternative of B3LYP for geometry optimizations of Rh-catalyzed C–H functionalization, which is proposed by Head-Gordon group in 2008 [50]. This functional includes empirical dispersion at DFT-D2 level, which gives a good accuracy of weak interaction. Moreover, the introduction of long-range correction into ωB97XD makes a good result in the calculation of charge transfer and Rydberg excitation. The time consumption of ωB97XD is also significantly higher than that of B3LYP [51, 52].

2.1.1 Basis Set

Generally, the time consuming of DFT calculations is mostly used in the calculation of double-election integral, which is positively correlated with N^4, in which N is the total Gaussian type functions for a specific molecule [53]. Therefore, the computation time consuming for a specific molecule depends on the selectivity of the basis set for all atoms in this molecule. In fact, the selection of the basis set is also arbitrary, which even could be based on experience and preferences [54]. Fortunately, the accuracy of most DFT methods is always independent of the size of the selected basis set, while most large enough basis sets can yield the same result in DFT calculations [55]. Additionally, when the number of base functions for using a basis set is the same, the time consuming can be saved by using a basis set with less Gaussian-type function under the same precision [56]. Therefore, segmented contraction basis sets, such as Pople's basis sets and def2 series of basis sets, are the better choice in DFT calculations [55, 57, 58].

For the computational study of Rh-catalyzed C–H functionalization, regular polarization functions are necessary, which can improve the accuracy to a great extent [59]. However, the larger angular momentum functions are unnecessary. A triple-zeta basis set is usually slightly better than a double-zeta one [60, 61]. Followed this idea, 6-31G(d) is the smallest acceptable basis set for the description of the atoms except for Rh metal in geometry optimizations [45, 62, 63]. The basis set of 6-311G(d,p) is another better choice for both accuracy and efficiency [50, 64]. When def2 series of basis set is used, def2-SVP is acceptable, while, def2-TZVP is a better one [65]. In an anionic molecule, defuse functions are necessary, therefore, 6-31 + G(d) is the acceptable basis set [44, 66]. To obtain accurate energy information, the polarization functions and defuse functions should be both taken into account in the energy calculations. The first choice of basis set for the description of the atoms except for Rh metal is 6-311 + G(d,p) followed by 6-31 + G(d) and 6-311G(d,p) in energy calculations [63, 67–69].

For the DFT calculation onto Rh metal atom, employing a pseudo potential basis set is strongly recommended for the consideration of both accuracy and efficiency. In this area, LANL2DZ [70] is the smallest basis set for Rh metal atom, which only provides only acceptable accuracy for some geometry optimizations and energy calculations [62, 64, 67, 69]. The larger ones, such as SDD [71], LANL2TZ [70], LANL08 [72], and LANL08(f) [73], are also recommended for Rh metal atom [63, 68].

2.1.2 Solvent Effect

The solvent effect is crucial to Rh-catalyzed C–H functionalization, which should always be considered in energy calculations. The solvent effect in homogeneous catalysis calculations is considered by the implicit solvent model. In the implicit solvent model, the solvent environment is simply considered as a polarizable continuous medium. The advantage of the implicit solvent model is that it can represent the average effect of solvents without the consideration of various possible molecular arrangements of solvent layer. The weakness is that the strong interaction between solvent and solute cannot be represented, such as hydrogen bond. Moreover, the accuracy of solvation energy for ionic solute cases is significantly lower than that of neutral solute cases.

Dielectric formulation (D-PCM) is the early member of PCM family, which only includes the charge density of the solute wavefunction within the solute surface into the solute–solvent interaction. The integral equation formalism PCM (IEF-PCM) includes the charge density of the wave function beyond the solute surface into the solute–solvent interaction [74]. The conductor-like PCM (C-PCM) developed is the implementation of conductor-like screening model in the PCM framework, which works well for solvents with a high dielectric constant such as water solvent [16, 75]. In isodensity PCM (I-PCM) and self-consistent IPCM (SCI-PCM), the solute cavity can be defined as a surface with a constant electron density (isodensity surface) [76]. The best implicit solvent model for DFT calculations at present may be SMD (solvation model based on density), where the bulk electrostatic contribution is calculated on the basis of the IEF-PCM protocol.

References

1. Bak KL, Hansen AE, Ruud K, Helgaker T, Olsen J, Jørgensen P (1995) Ab initio calculation of electronic circular dichroism for trans-cyclooctene using London atomic orbitals. Theor Chim Acta 90:441–458
2. Abegg PW (1975) Ab initiocalculation of spin-orbit coupling constants for gaussian lobe and gaussian-type wave functions. Mol Phys 30:579–596
3. Andrae D, Häußermann U, Dolg M, Stoll H, Preuß H (1990) Energy-adjustedab initio pseudopotentials for the second and third row transition elements: Molecular test for M2 (M = Ag, Au) and MH (M = Ru, Os). Theor Chim Acta 77:123–141
4. Almlöf J, Faegri K, Korsell K (1982) Principles for a direct SCF approach to LCAO-MO Ab-Initio calculations. J Comput Chem 3:385–399
5. Schrödinger E (1926) Quantisierung als Eigenwert problem. Ann Phys 384:361–376
6. Stewart JJP (1989) Optimization of parameters for semiempirical methods II Applications. J Comput Chem 10:221–264
7. Daniels AD, Millam JM, Scuseria GE (1997) Semiempirical methods with conjugate gradient density matrix search to replace diagonalization for molecular systems containing thousands of atoms. J Chem Phys 107:425–431
8. Salahub DR, Zerner MC (1989) The Challenge of d and f Electrons. ACS, Washington, D.C.
9. Kohn W, Sham LJ (1965) Self-consistent equations including exchange and correlation effects. Phys Rev 140:A1133–A1138
10. Lewars EG (2011) Computational chemistry. Springer, Dordrecht, Peterborough, pp 45–83
11. Perdew JP, Schmidt K (2001) Density functional theory and its application to materials. AIP Melville, New York
12. Zhao Y, Truhlar DG (2006) A new local density functional for main-group thermochemistry, transition metal bonding, thermochemical kinetics, and noncovalent interactions. J Chem Phys 125:194101–1941017
13. Tao J, Perdew JP, Staroverov VN, Scuseria GE (2003) Climbing the density functional ladder: nonempirical meta-generalized gradient approximation designed for molecules and solids. Phys Rev Lett 91:146401–146404
14. Van Voorhis T, Scuseria GE (1998) A novel form for the exchange-correlation energy functional. J Chem Phys 109:400–410
15. Becke AD (1993) Density-functional thermochemistry. III. The role of exact exchange. J Chem Phys 98:5648–5652
16. Barone V, Cossi M (1998) Quantum calculation of molecular energies and energy gradients in solution by a conductor solvent model. J Phys Chem A 102:1995–2001
17. Becke AD (1997) Density-functional thermochemistry. V. Systematic optimization of exchange-correlation functionals. J Chem Phys 107:8554–8560
18. Schmider HL, Becke AD (1998) Optimized density functionals from the extended G2 test set. J Chem Phys 108:9624–9631

19. Cohen AJ, Handy NC (2001) Dynamic correlation. Mol Phys 99:607–615
20. Perdew JP, Burke K, Ernzerhof M (1996) Generalized gradient approximation made simple. Phys Rev Lett 77:3865–3868
21. Adamo C, Barone V (1999) Toward reliable density functional methods without adjustable parameters: The PBE0 model. J Chem Phys 110:6158–6170
22. Adamo C, Barone V (1998) Exchange functionals with improved long-range behavior and adiabatic connection methods without adjustable parameters: The mPW and mPW1PW models. J Chem Phys 108:664–675
23. Xu X, Goddard WA III (2004) From the cover: The X3LYP extended density functional for accurate descriptions of nonbond interactions, spin states, and thermochemical properties. PANS 101:2673–2677
24. Zhao Y, Schultz NE, Truhlar DG (2005) Exchange-correlation functional with broad accuracy for metallic and nonmetallic compounds, kinetics, and noncovalent interactions. J Chem Phys 123:161103
25. Zhao Y, Truhlar DG (2006) Comparative DFT study of van der Waals complexes: rare-gas dimers, alkaline-earth dimers, zinc dimer, and zinc-rare-gas dimers. J Phys Chem A 110:5121–5129
26. Zhao Y, Truhlar DG (2007) The M06 suite of density functionals for main group thermochemistry, thermochemical kinetics, noncovalent interactions, excited states, and transition elements: two new functionals and systematic testing of four M06-class functionals and 12 other functionals. Theor Chem Acc 120:215–241
27. Zhao Y, Truhlar DG (2006) Density functional for spectroscopy: no long-range self-interaction error, good performance for Rydberg and charge-transfer states, and better performance on average than B3LYP for ground states. J Phys Chem A 110:13126–13130
28. Lynch BJ, Fast PL, Harris M, Truhlar DG (2000) Adiabatic connection for kinetics. J Phys Chem A 104:4811–4815
29. Lingwood M, Hammond JR, Hrovat DA, Mayer JM, Borden WT (2006) MPW1K performs much better than B3LYP in DFT calculations on reactions that proceed by proton-coupled electron transfer (PCET). J Chem Theory Comput 2:740–745
30. Grimme S (2006) Semiempirical hybrid density functional with perturbative second-order correlation. J Chem Phys 124:034108–034123
31. Schwabe T, Grimme S (2006) Towards chemical accuracy for the thermodynamics of large molecules: new hybrid density functionals including non-local correlation effects. Phys Chem Chem Phys 8:4398–4401
32. Head-Gordon M, Pople JA, Frisch MJ (1988) MP2 energy evaluation by direct methods. Chem Phys Lett 153:503–506
33. Cremer C (2002) Essentials of computational chemistry. Wiley, Chichester, United Kingdom
34. Niu S, Hall MB (2000) Theoretical studies on reactions of transition-metal complexes. Chem Rev 100:353–406
35. Becke AD (1988) Density-functional exchange-energy approximation with correct asymptotic behavior. Phys Rev A 38:3098–3100
36. Lee C, Yang W, Parr RG (1988) Development of the Colle-Salvetti correlation-energy formula into a functional of the electron density. Phys Rev B 37:785–789
37. Wiedemann SH, Lewis JC, Ellman JA, Bergman RG (2006) Experimental and computational studies on the mechanism of N-heterocycle C–H activation by Rh(I). J Am Chem Soc 128:2452–2462
38. Mai BK, Szabó KJ, Himo F (2018) Mechanisms of Rh-Catalyzed oxyfluorination and oxytrifluoromethylation of diazocarbonyl compounds with hypervalent fluoroiodine. ACS Catal 8:4483–4492
39. Cui C-X, Xu D, Ding B-W, Qu L-B, Zhang Y-P, Lan Y (2019) Benchmark study of popular density functionals for calculating binding energies of three-center two-electron bonds. J Comput Chem 40:657–670
40. Zhao Y, Truhlar DG (2008) Density functionals with broad applicability in chemistry. Acc Chem Res 41:157–167

41. Taylor DE, Angyan JG, Galli G, Zhang C, Gygi F, Hirao K, Song JW, Rahul K, Anatole von Lilienfeld O, Podeszwa R, Bulik IW, Henderson TM, Scuseria GE, Toulouse J, Peverati R, Truhlar DG, Szalewicz K (2016) Blind test of density-functional-based methods on intermolecular interaction energies. J Chem Phys 145:124105

42. Grimme S, Ehrlich S, Goerigk L (2011) Effect of the damping function in dispersion corrected density functional theory. J Comput Chem 32:1456–1465

43. Yanai T, Tew DP, Handy NC (2004) A new hybrid exchange–correlation functional using the Coulomb-attenuating method (CAM-B3LYP). Chem Phys Lett 393:51–57

44. Wang X, Gensch T, Lerchen A, Daniliuc CG, Glorius F (2017) Cp*Rh(III)/bicyclic olefin cocatalyzed C–H bond amidation by intramolecular amide transfer. J Am Chem Soc 139:6506–6512

45. Neufeldt SR, Jimenez-Oses G, Huckins JR, Thiel OR, Houk KN (2015) Pyridine N-Oxide vs pyridine substrates for Rh(III)-Catalyzed oxidative C–H bond functionalization. J Am Chem Soc 137:9843–9854

46. Mardirossian N, Head-Gordon M (2016) How accurate are the minnesota density functionals for noncovalent interactions, isomerization energies, thermochemistry, and barrier heights involving molecules composed of main-group elements? J Chem Theory Comput 12:4303–4325

47. Peverati R, Truhlar DG (2011) M11-L: a local density functional that provides improved accuracy for electronic structure calculations in chemistry and physics. J Phys Chem Lett 3:117–124

48. Yu S, Tang G, Li Y, Zhou X, Lan Y, Li X (2016) Anthranil: an aminating reagent leading to bifunctionality for both C(sp(3))-H and C(sp(2))-H under Rhodium(III) Catalysis. Angew Chem Int Ed 55:8696–8700

49. Lin Y-S, Li G-D, Mao S-P, Chai J-D (2013) Long-range corrected hybrid density functionals with improved dispersion corrections. J Chem Theory Comput 9:263–272

50. Zheng C, Zheng J, You S-L (2015) A DFT study on Rh-Catalyzed asymmetric dearomatization of 2-Naphthols initiated with C–H activation: a refined reaction mechanism and origins of multiple selectivity. ACS Catal 6:262–271

51. Kupka T, Nieradka M, Stachow M, Pluta T, Nowak P, Kjaer H, Kongsted J, Kaminsky J (2012) Basis set convergence of indirect spin-spin coupling constants in the Kohn-Sham limit for several small molecules. J Phys Chem A 116:3728–3738

52. Roca-Sabio A, Regueiro-Figueroa M, Esteban-Gómez D, de Blas A, Rodríguez-Blas T, Platas-Iglesias C (2012) Density functional dependence of molecular geometries in lanthanide(III) complexes relevant to bioanalytical and biomedical applications. Comput Theor Chem 999:93–104

53. Ditchfield R, Hehre WJ, Pople JA (1971) Self-consistent molecular-orbital methods. ix. an extended gaussian-type basis for molecular-orbital studies of organic molecules. J Chem Phys 54:724–728

54. Heaton-Burgess T, Yang W (2008) Optimized effective potentials from arbitrary basis sets. J Chem Phys 129:194102–194112

55. Weigend F (2006) Accurate Coulomb-fitting basis sets for H to Rn. Phys Chem Chem Phys 8:1057–1065

56. Huang J, Kertesz M (2004) Intermolecular transfer integrals for organic molecular materials: can basis set convergence be achieved? Chem Phys Lett 390:110–115

57. Xu X, Truhlar DG (2011) Accuracy of effective core potentials and basis sets for density functional calculations, including relativistic effects, as illustrated by calculations on arsenic compounds. J Chem Theory Comput 7:2766–2779

58. Jensen F (2015) Segmented contracted basis sets optimized for nuclear magnetic shielding. J Chem Theory Comput 11:132–138

59. Eichkorn K, Treutler O, Öhm H, Häser M, Ahlrichs R (1995) Auxiliary basis sets to approximate Coulomb potentials. Chem Phys Lett 240:283–290

60. Almlöf J, Taylor PR (1987) General contraction of Gaussian basis sets. I. Atomic natural orbitals for first- and second- row atoms. J Chem Phys 86:4070–4077

61. Dyall KG (2016) Relativistic double-zeta, triple-zeta, and quadruple-zeta basis sets for the light elements H–Ar. Theor Chem Acc 135
62. Hawkes KJ, Cavell KJ, Yates BF (2008) Rhodium-Catalyzed C–C coupling reactions: mechanistic considerations. Organometallics 27:4758–4771
63. Zhao D, Li X, Han K, Li X, Wang Y (2015) Theoretical investigations on Rh(III)-catalyzed cross-dehydrogenative aryl-aryl coupling via C–H bond activation. J Phys Chem A 119:2989–2997
64. Santoro S, Himo F (2018) Mechanism and selectivity of rhodium-catalyzed C–H bond arylation of indoles. Int J Quantum Chem 118:e25526
65. Lied F, Lerchen A, Knecht T, Mück-Lichtenfeld C, Glorius F (2016) Versatile Cp*Rh(III)-Catalyzed selective Ortho-chlorination of arenes and heteroarenes. ACS Catal 6:7839–7843
66. Yu S, Liu S, Lan Y, Wan B, Li X (2015) Rhodium-catalyzed C–H activation of phenacyl ammonium salts assisted by an oxidizing C–N bond: a combination of experimental and theoretical studies. J Am Chem Soc 137:1623–1631
67. Tran G, Hesp KD, Mascitti V, Ellman JA (2017) Base-controlled completely selective linear or branched Rhodium(I)-Catalyzed C–H ortho-Alkylation of azines without preactivation. Angew Chem Int Ed 56:5899–5903
68. Dang Y, Qu S, Tao Y, Deng X, Wang Z-X (2015) Mechanistic insight into ketone alpha-alkylation with unactivated Olefins via C–H activation promoted by metal-organic cooperative catalysis (MOCC): enriching the MOCC Chemistry. J Am Chem Soc 137:6279–6291
69. Yamaguchi T, Natsui S, Shibata K, Yamazaki K, Rej S, Ano Y, Chatani N (2019) Rhodium-Catalyzed alkylation of C–H bonds in aromatic amides with non-activated 1-Alkenes: the possible generation of carbene intermediates from alkenes. Chem Eur J 25:6915–6919
70. Hay PJ, Wadt WR (1985) Ab initio effective core potentials for molecular calculations. Potentials for K to Au including the outermost core orbitals. J Chem Phys 82:299–310
71. Fuentealba P, Preuss H, Stoll H, Von Szentpály L (1982) A proper account of core-polarization with pseudopotentials: single valence-electron alkali compounds. Chem Phys Lett 89:418–422
72. Roy LE, Hay PJ, Martin RL (2008) Revised basis sets for the LANL effective core potentials. J Chem Theory Comput 4:1029–1031
73. Ehlers AW, Böhme M, Dapprich S, Gobbi A, Höllwarth A, Jonas V, Köhler KF, Stegmann R, Veldkamp A, Frenking G (1993) A set of f-polarization functions for pseudo-potential basis sets of the transition metals Sc–Cu, Y–Ag and La–Au. Chem Phys Lett 208:111–114
74. Tomasi J, Mennucci B, Cancès E (1999) The IEF version of the PCM solvation method: an overview of a new method addressed to study molecular solutes at the QM ab initio level. J Mol Struc (THEOCHEM) 464:211–226
75. Cossi M, Rega N, Scalmani G, Barone V (2003) Energies, structures, and electronic properties of molecules in solution with the C-PCM solvation model. J Comput Chem 24:669–681
76. Foresman JB, Keith TA, Wiberg KB, Snoonian J, Frisch MJ (1996) Solvent effects: influence of cavity shape, truncation of electrostatics, and electron correlation on ab initio reaction field calculations. J Phys Chem 100:16098–16104

Chapter 3
Theoretical Study of Rh-Catalyzed C–C Bond Formation Through C–H Activation

Song Liu, Cheng-Xing Cui, Ruopeng Bai, Chun-Xiang Li, and Yu Lan

Construction of C–C bonds is a considerable area in both academic and industrial fields [1, 2]. The ubiquity and low cost of hydrocarbons make C–H bond functionalization to be an attractive alternative to classical transition metal-catalyzed cross-coupling reactions with functionalized organic compounds [2–7]. In the last few decades, numerous synthetic methods have been devoted to form C–C bonds and lengthen carbon chains through C–H bond functionalization. It's worth noting that the introduction of transition metal makes C–H bond functionalization to be a facile process. Among these transition metals, Rh [8–16] used as catalysts stands out for their functional group tolerance and wide range of synthetic utility in the C–C bond formation reactions with C–H bonds. In a series of pioneering works, a variety of Rh-catalyzed C–C coupling reactions with C–H bonds have been reported [17, 18]. Because of the complexity of the C–H bond activation reaction itself and its mechanism, the computational study processes have witnessed tremendous development to aid in the design of Rh-catalyzed C–C coupling reactions with C–H bonds [19, 20]. In particular, the theoretical studies of Rh-catalyzed C–H bond arylation, alkylation, vinylation, alkynylation, carbonylation, and annulation have been of interest, which would be discussed in detail.

The C–H bonds can be considered nucleophile because the electronegativity of carbon is higher than that of hydrogen in this moiety. Therefore, the C–H bond can formally donate both of their bonding electrons to an electrophile to form a new C–C σ-bond through redox-neutral cross-coupling. Alternatively, C–H bonds also can couple with a nucleophile in the presence of exogenous oxidants, which results in oxidative-coupling reactions (Scheme 3.1). In this case, the exogenous oxidant removes one pair of electrons from C–H bond and another nucleophile to build a new C–C σ-bond.

© The Author(s), under exclusive license to Springer Nature Singapore Pte Ltd. 2021
Y. Lan et al., *Computational Advances of Rh-Catalyzed C–H Functionalization*,
SpringerBriefs in Molecular Science,
https://doi.org/10.1007/978-981-16-0432-4_3

Scheme 3.1 Rh-catalyzed
coupling reaction with C–H
bonds

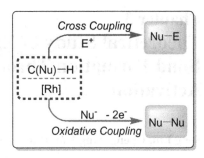

3.1 Rh-Catalyzed C–H Bond Arylation

Generally, the Rh-catalyzed C–H bond arylations can be categorized as redox-neutral
cross-coupling reactions with electrophilic aryl reagents and oxidative-coupling
reactions with nucleophilic aryl reagents.

3.1.1 C–H Bond Arylation with Electrophiles

In Rh-catalyzed C–H bond arylation reactions through redox-neutral cross-coupling,
aryl bromide and aryl iodide [21] are often used as electrophiles to coupling with
C–H bond. The common redox-type catalytic cycle for the Rh-catalyzed C–H bond
arylation with electrophilic aryl reagents is concluded in Scheme 3.2, which majorly
involves oxidative addition with aryl halides, base-assisted C–H bond cleavage, and
reductive elimination.

As the earliest example, Bedford and co-workers [22] reported a Rh(I)-catalyzed
intermolecular *ortho*-arylation of phenols with aryl halides (Scheme 3.3a). This efficient reaction proceeded with Rh(I)ClP(Ph₃)₃ as catalyst in the presence of the corresponding phosphinites as co-catalysts, which needs to be prepared for each individual

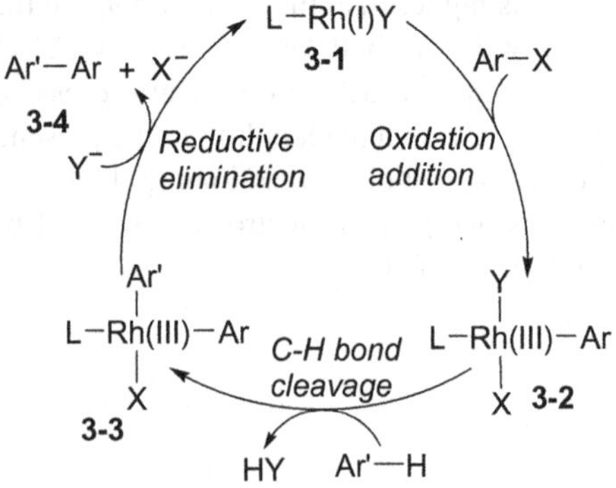

Scheme 3.2 The common
redox-type catalytic cycle for
Rh-catalyzed C–H bond
arylation with electrophiles

$$L-Rh(I)Y$$
$$\mathbf{3\text{-}1}$$

$$Ar'-Ar + X^-$$
$$\mathbf{3\text{-}4}$$
$$Y^-$$

Reductive
elimination

$$Ar-X$$

Oxidation
addition

$$\begin{array}{c} Ar' \\ | \\ L-Rh(III)-Ar \\ | \\ X \end{array}$$
$$\mathbf{3\text{-}3}$$

C-H bond
cleavage

$$\begin{array}{c} Y \\ | \\ L-Rh(III)-Ar \\ | \\ X \end{array} \mathbf{3\text{-}2}$$

$$HY \qquad Ar'-H$$

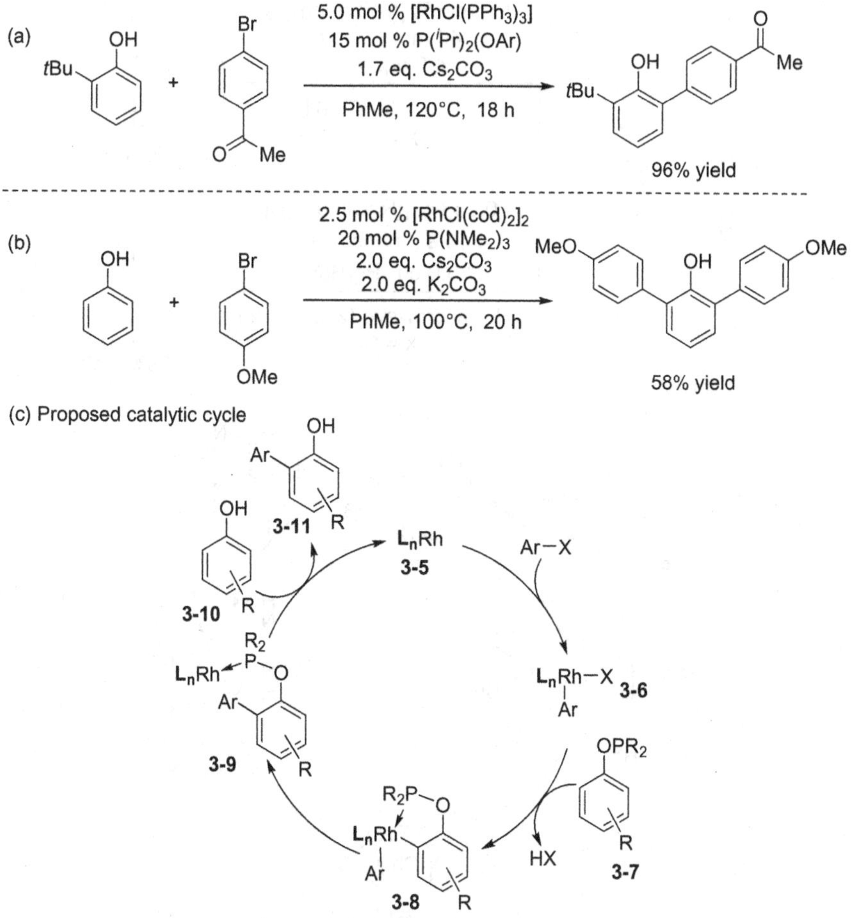
(a)
OH
tBu
Br
O
Me
5.0 mol % [RhCl(PPh3)3]
15 mol % P(iPr)2(OAr)
1.7 eq. Cs2CO3
PhMe, 120°C, 18 h
96% yield
(b)
2.5 mol % [RhCl(cod)2]2
20 mol % P(NMe2)3
2.0 eq. Cs2CO3
2.0 eq. K2CO3
PhMe, 100°C, 20 h
MeO
OMe
58% yield
(c) Proposed catalytic cycle
3-11
LnRh
3-5
Ar–X
3-10
LnRh–X
3-6
3-9
OPR2
3-7
HX
3-8

Scheme 3.3 Rh(I)-catalyzed arene arylation with electrophiles and proposed catalytic cycle

substrate prior to catalysis. This limitation was also addressed by the same group by using inexpensive $P(NMe_2)_3$ as the co-catalyst (Scheme 3.3b). The authors proposed that the phosphinite co-catalyst coordinates onto the Rh(III) center, which is formed by oxidative addition of the aryl halide (Scheme 3.3c). Subsequent base-assisted C–H bond cleavage gives the aryl-Rh intermediate **3-8**, which undergoes reductive elimination to lead to the generation of the 2-arylated aryl dialkylphosphinite coordinated Rh(I) complex **3-9**. The catalytic transesterification of 2-arylated aryl dialkylphosphinite with phenol regenerates the co-catalyst and liberates the 2-arylated phenol product **3-11**.

In another example, Ellman and Bergman have developed a series of intermolecular arylation of azoles with aryl halides through Rh-catalyzed C–H bond activation [23]. As shown in Scheme 3.4, aryl iodide and aryl bromine were employed as elec-

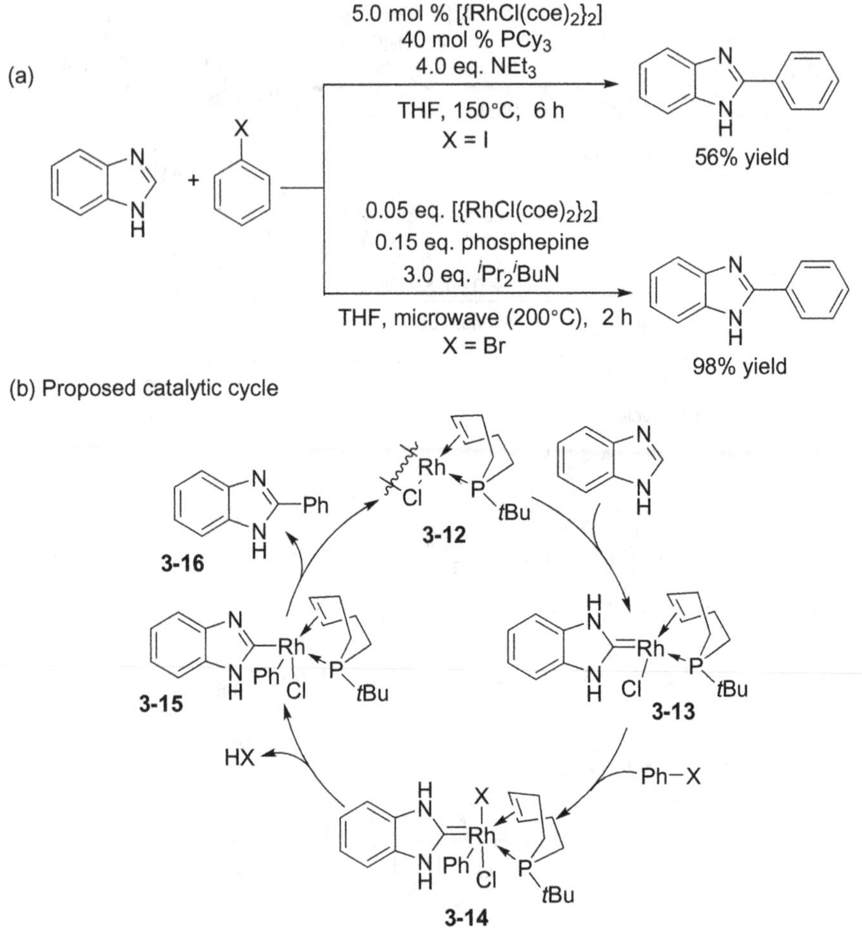
(a)
5.0 mol % [{RhCl(coe)2}2]
40 mol % PCy3
4.0 eq. NEt3
THF, 150°C, 6 h
X = I
56% yield
0.05 eq. [{RhCl(coe)2}2]
0.15 eq. phosphepine
3.0 eq. iPr2iBuN
THF, microwave (200°C), 2 h
X = Br
98% yield
(b) Proposed catalytic cycle
3-12
3-13
3-14
3-15
3-16
Ph−X
HX

Scheme 3.4 Rh(I)-catalyzed intermolecular arylation of nitrogen heterocycles with aryl halides and proposed catalytic cycle

trophiles in this reaction. The more rigorous reaction condition of aryl bromine with benzimidazole indicated that the oxidative addition might be the rate limit for the catalytic cycle.

Based on detailed mechanistic studies, a redox mechanism was proposed (Scheme 3.4b). The arylation reaction starts with the initial coordination of benzimidazole onto Rh(I) followed by a 1,2-hydrogen shift generating an *N*-heterocyclic carbene rhodium complex **3-13**, which plays the key intermediate in the catalytic cycle. The oxidative addition of aryl halides onto the Rh(I) center in the Rh–carbene complex **3-13** leads to the formation of aryl Rh(III)-carbene intermediate **3-13**. With the assistance of base, deprotonation of **3-14** gives the diaryl-Rh(III) complex **3-15**, which undergoes reductive elimination to yield the arylation product **3-16** and regenerate the Rh(I) species **3-12**.

In 2005, Sames and co-workers [24] reported a [Rh(coe)$_2$Cl]$_2$ catalyzed C2-selective arylation of indoles in the presence of an electron-poor phosphine ligand [(*p*-CF$_3$-C$_6$H$_4$)$_3$P] and cesium pivalate by using aryl iodides as electrophile (Scheme 3.5). The reaction temperature is optimized as 120 °C, which indicates a high observable activation free energy. In an experiment, a five-coordinated Rh(III) species **3-17**, which was isolated and characterized by ^{31}P NMR analysis, can react with indole to furnish 2-phenyl indole product. This observation indicated complex **3-17** was reasonably considered to be the active species in the catalytic cycle.

Santoro and co-worker [25] performed a DFT study to investigate the mechanism of the above arylation reaction. As shown in Fig. 3.6, the catalytic cycle starts from the four-coordinated anionic Rh(I) complex **3-27**, in which the phenyl iodide coordinates with Rh(I). The oxidative addition of phenyl iodide onto Rh(I) occurs via transition state **3-28ts** to generate a five-coordinated phenyl Rh(III) intermediate **3-29**. With the coordination of another ligand and releasing an iodide, the active species **3-18** is formed. The following step is the release of one phosphine ligand with the

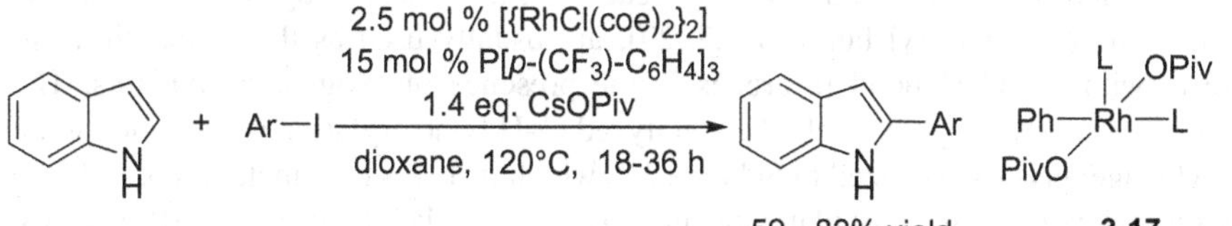

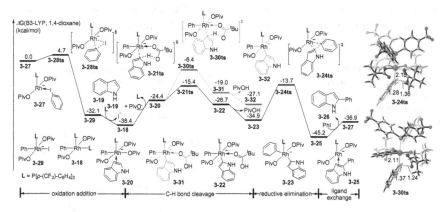

Scheme 3.5 Rh(I)-catalyzed intermolecular arylation of indoles with aryl halides

Fig. 3.6 Free energy profiles for the Rh(I)-catalyzed o C2-selective C–H bond activation and arylation of indoles. The values are the relative free energies given in kcal/mol calculated at the B3-LYP/6-31G(d,p)/LANL2DZ//B3-LYP/6-311+G(d,p)/LANL2DZ level of theory in bromobenzene

coordination of the additional pivalate and indole, which is 14.0 kcal/mol endogenic. The CMD type C–H bond cleavage at C2 position occurs via transition state **3-21ts** generating aryl rhodium intermediate **3-22** with an energy barrier of 23.0 kcal/mol. As a contrast, the corresponding energy barrier for the C–H cleavage at C3 position via transition state **3-30ts** is 32.0 kcal/mol. The relative energy of **3-30ts** is 9.0 kcal/mol higher than that of transition state **3-21ts**, which is in agreement with the experimental results that the C–H activation at C2 is more favorable than that at C3 position. The comparison of the geometry information of **3-21ts** and **3-30ts** shows that **3-21ts** is an earlier transition state, which is a consequence of the higher nucleophilicity of C3 and donates the lower activation free energy of transition state **3-21ts**. Then, the reductive elimination of diaryl-Rh(III) complex **3-23** gives the C2-arylated indole intermediate **3-25** via transition state **3-24ts** with an overall activation free energy of 24.7 kcal/mol. After a release of product **3-26** and coordination of phenyl iodide, the Rh(I) complex **3-27** is regenerated. The calculated results indicate that reductive elimination is the rate-determining step in the catalytic cycle.

3.1.1.1 C–H Bond Arylation with Another Nucleophile

In Rh-catalyzed C–H bond arylation reactions, aryl metal reagents, such as aryl stannum [26], and aryl boranes [27–29], are usually used as the nucleophiles to react with the C–H bond in arenes in the presence of exogenous oxidants. The common catalytic cycle for the Rh-catalyzed C–H bond arylation with nucleophilic aryl reagents is summarized in Scheme 3.7, which involves transmetallation of Rh(I) with aryl metal reagent, oxidated by an exogenous oxidant to afford Rh(III) species, base-assisted C–H bond cleavage, and reductive elimination.

As shown in Scheme 3.8a, Oi and co-workers [26] reported the first example of direct arylation of pyridinyl arene with stannanes by using Wilkinson catalyst [Rh(I)Cl(PPh₃)₃] in 1, 2-dichloroethane solvent. This reaction can give both mono- and bis-arylation products in yields of 56 and 20%, respectively. Although the

Scheme 3.7 Common mechanism of Rh-catalyzed C–H bond arylation with nucleophiles starting from a Rh(I) species

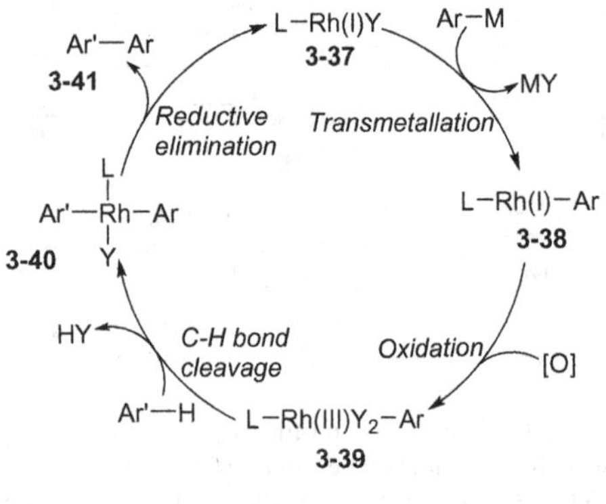

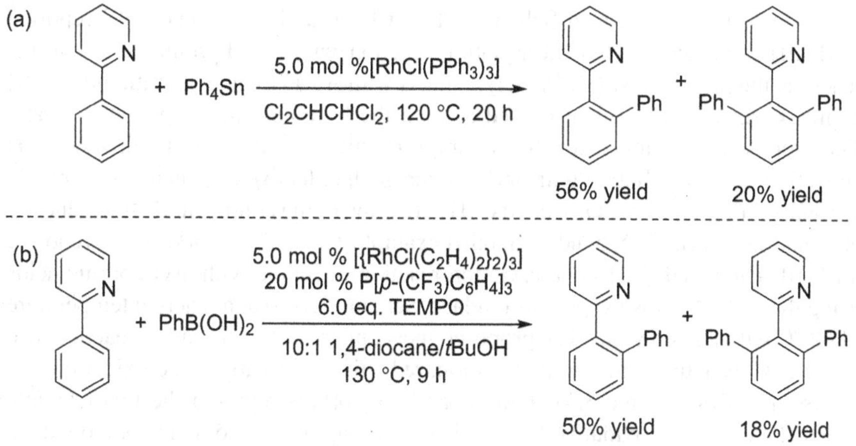

(a)

56% yield   20% yield

(b)

50% yield   18% yield

Scheme 3.8 Rh(I)-catalyzed arene oxidative arylation with nucleophiles

detailed data on the mechanism of the reaction were not clear, the 1, 2-dichloroethane solvent was considered to serve as the exogenous oxidant in this transformation.

The 2,2,6,6-tetramethylpiperidinyloxyl (TEMPO) also can be used as the exogenous oxidant in Rh-catalyzed C–H bond arylation reactions. As shown in Scheme 3.8b, Studer and co-workers [29] reported a Rh(I)-catalyzed C–H bond arylation of arenes with arylboronic acids using TEMPO as the exogenous oxidant to keep the redox neutral. The reaction occurs at 100 °C with acceptable yields of both mono- and bis-arylation products.

The sequence of these elementary reactions can be swapped in Rh-catalyzed C–H bond arylation reactions with nucleophilic aryl reagents. When using high-valent Rh species as the active catalyst, the base-assisted C–H bond cleavage can be the first step in the catalytic cycle, while, the oxidation of Rh(I) to Rh(III) can be the last step to regenerate active catalyst and keep redox neutral. In this case, the common catalytic cycle involves base-assisted C–H bond cleavage, transmetallation with aryl metal reagent, reductive elimination, and oxidation by an exogenous oxidant (Scheme 3.9).

Scheme 3.9 Common mechanism of Rh-catalyzed C–H bond arylation with nucleophiles starting from a Rh(III) species

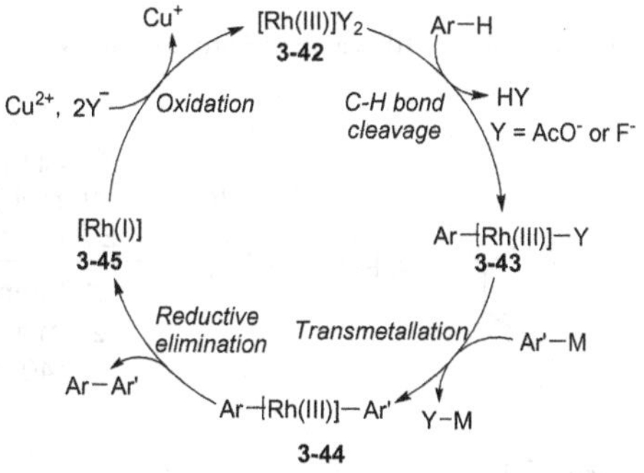

As an example shown in Scheme 3.10a, Cheng and co-workers [30] reported a Rh(III)-catalyzed C–H bond arylation reaction with aryl boronic acids. In this reaction, the pre-catalyst {RhCp*Cl$_2$}$_2$ reacts with Ag$_2$O to give the cationic Rh(III) complex. Excess Ag$_2$O is also used as both the base and the exogenous oxidant. The proposed mechanism for this arylation reaction involves amido-directed C–H activation, transmetallation with aryl boronic acid, reductive elimination to form C–C bond, oxidation by Ag$_2$O, second C–H activation of the coming aryl, and reductive elimination to form C–N bond. In another example, Cui and co-workers [31] reported a Rh(III)-catalyzed C–H bond arylation reaction of indole with aryl boronic acids using the Cu(OAc)$_2$ as exogenous oxidant (Scheme 3.10b) at a reaction temperature of 60 °C with a good yield. The proposed mechanism for this arylation reaction also involves C–H activation, transmetallation, reductive elimination, and oxidation.

As shown in Scheme 3.11, Glorius and co-workers reported the first example of Rh(III)-catalyzed oxidative C–H/C–H cross-coupling of aromatic compounds in 2012 [32]. In this reaction, benzamides smoothly reacted with solvent amounts of haloarenes to give the desired products in good yields. In addition, halides, including bromides, chlorides, and iodides, were all compatible with this Rh-catalysis system.

Zhao and co-workers [33] performed a DFT calculation to investigate the mechanism of the Rh(III)-catalyzed oxidative C–H/C–H cross-coupling reaction. The calculated free energy profile for the rational reaction pathway is shown in Fig. 3.12.

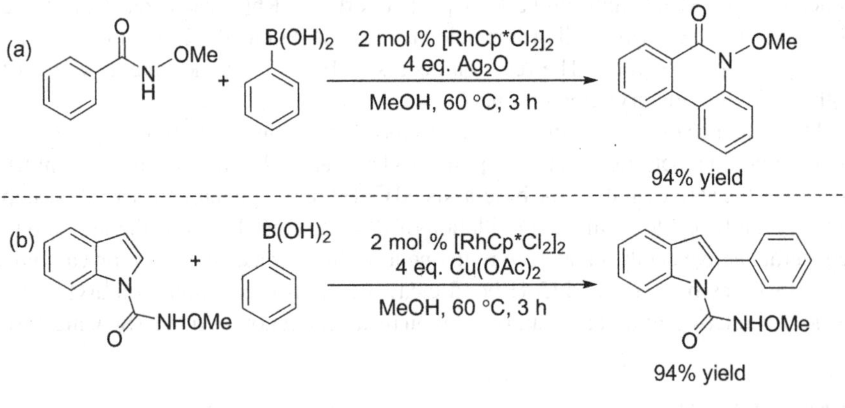

Scheme 3.10 Rh(III)-catalyzed arene oxidative arylation with nucleophiles

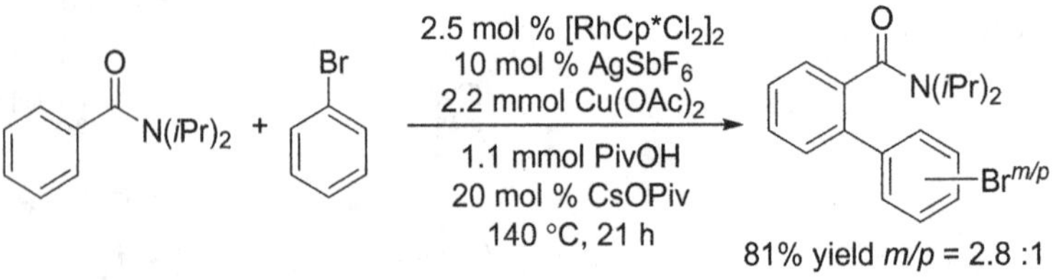

Scheme 3.11 Rh(III)-catalyzed oxidative C–H/C–H cross-coupling of aromatic compounds

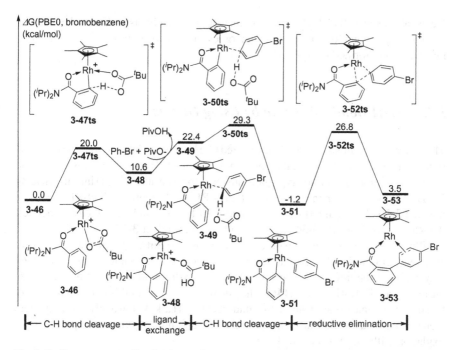

Fig. 3.12 Free energy profiles for the Rh(III)-catalyzed oxidative C–H/C–H cross-coupling reaction. The values are the relative free energies given in kcal/mol calculated at the PBE0/6-31G(d)/SDD//PBE0/6-311+G(d,p)/SDD level of theory in bromobenzene

The amido directed C–H bond activation occurs via a CMD type transition state **3-47ts**, in which the acetate acts as the base to deprotonate the ortho aromatic proton, with concomitant formation of an aryl-Rh(III) intermediate **3-48**. The activation barrier for this C–H bond cleavage via transition state **3-47ts** is 20.0 kcal/mol. After ligand exchange of PivO⁻ with PivOH, the subsequent C–H activation of aryl bromide occurs via transition state **3-50ts**, in which the PivO⁻ also acts as the base to conduct the deprotonation. This electrophilic deprotonative metalation is calculated to be the rate-determining step in the catalytic cycle with an overall activation free energy of 29.3 kcal/mol. The C(aryl)-C(aryl) reductive elimination of diaryl-Rh(III) intermediate **3-51** gives the arylation product coordinated Rh(I) complex **3-53**. The activation barrier for the reductive elimination step is 28.0 kcal/mol, which indicated that reductive elimination is also a slow process.

3.2 Rh-Catalyzed C–H Bond Alkylation

Rh-catalyzed C–H bond alkylation is an efficient way to construct $C(sp^2)$-$C(sp^3)$ bond [13, 17, 34–37]. Olefins [38–41], alkynes [8, 42, 43], and carbenoids [44, 45]

are usually used as the alkyl source in this reaction. Mechanistically, the key step of $C(sp^2)$–$C(sp^3)$ bond formation often undergoes an insertion of unsaturated molecules into the newly formed C(aryl)–Rh bond.

3.2.1   C–H Bond Alkylation by Using Olefins

The general mechanism for the Rh-catalyzed C–H bond alkylation of olefins is shown in Scheme 3.13. The C–H bond activation by using Rh(I) species **3-54** would occur through an oxidative addition type mechanism to form an aryl-Rh(III) intermediate **3-55**. The following olefin insertion into C(aryl)–Rh bond generates alkyl-Rh(III) intermediate **3-56**. The final reductive elimination of hydride and alkyl then regenerates the Rh(I) active catalyst **3-54** and yields alkylation product **3-57**.

Ellman, Bergman, and co-workers have made a significant contribution in Rh-catalyzed intramolecular alkylation of heterocycle using simple olefins. In 2006, they reported the intermolecular alkylation of dihydroquinazolines with alkenes in the presence of a Rh(I) catalyst [46]. The kinetic monitoring of this reaction indicated that dihydroquinazoline substrate first undergoes C–H alkylation to give the alkylated arene, which was further aromatized to the corresponding quinazoline products by oxidation with MnO_2 (Scheme 3.14).

The offhandedly computational studies on the mechanism of dihydroquinazoline C–H bond activation were reported by the same group [47]. As shown in Fig. 3.15, the DFT calculations revealed that the C–H bond activation is initiated by the coordination of sp^2 nitrogen in dihydropyrimidine onto Rh(I) in species **3-58**. The isomerization of the intermediate **3-58** occurs via transition state **3-59ts** with an activation barrier of 23.2 kcal/mol to give the C–H σ-complex **3-60**. The subsequent C–H bond activation of dihydropyrimidine occurs through an oxidative addition-type mechanism via three-membered ring-type transition state **3-61ts** to provide the aryl Rh(III)-hydride species **3-62**. Then an intramolecular reductive hydrogen shift from Rh to

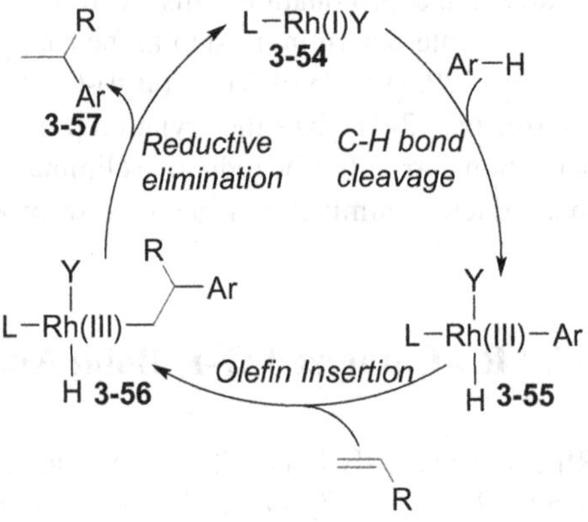

Scheme 3.13 General mechanism for the Rh-catalyzed C–H bond alkylation of olefins

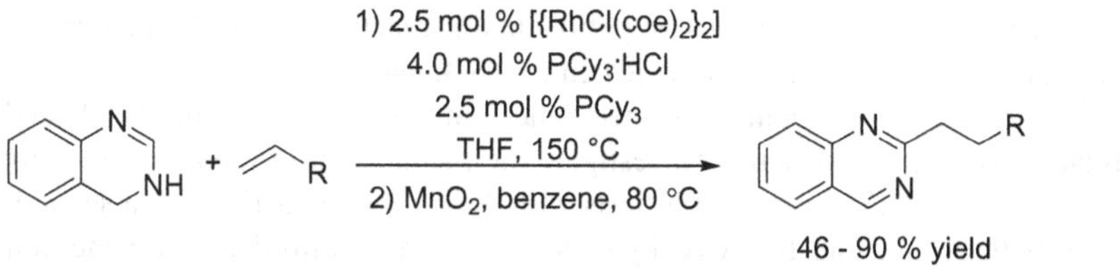

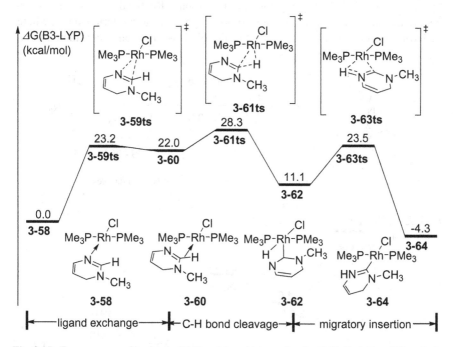

Scheme 3.14 Rh(I)-catalyzed intermolecular C–H alkylation of 3,4-dihydroquinazolines with olefins

Fig. 3.15 Free energy profiles for the Rh(I)-catalyzed intramolecular C–H alkylation of 3-methyl-3,4-dihydropyrimidine. The values are the relative free energies given in kcal/mol calculated at the B3-LYP/LACVP**++//B3-LYP/LACVP** level of theory

dihydropyrimidinyl in this species occurs via a four-membered ring-type transition state **3-63ts** resulting a Rh(I)-NHC complex **3-64**, which can further intermolecularly react with olefins to accomplish alkylation. The carbenation of Rh(I) was considered to be exergonic by 4.3 kcal/mol at 298 K.

In 2001, Bergman, Ellman, and co-workers reported an intramolecular C–H alkylation of imidazoles [48]. In this reaction, PCy₃ was proved to be an optimal ligand with [RhCl(coe)₂]₂ as the catalyst precursor to get high yields. The second position of the imidazoles was alkylated from five- or six-membered rings undergoing catalytic C–H bond activation. Moreover, a range of substrates including mono-, di-, and trisubstituted alkenyls are compatible in this reaction and allow the formation of alkylation product with stereogenic centers (Scheme 3.16).

In 2002, the same team performed a mechanistic study using imidazole and PH$_3$ ligands as a model system to investigate the mechanism of the above reaction. As shown in Fig. 3.17, this reaction starts with the C–H bond activation at second position of the imidazole followed by proton transfer to afford Rh–carbene complex **3-65**, which is the key intermediate for the subsequent transformation. The detailed mechanism of this process is given in Fig. 3.17. When Rh–carbene complex **3-65** is formed, DFT calculation indicated that the subsequent intramolecular metathesis-type insertion of alkenyl group into the C(carbene)-Rh bond occurs via a four-membered ring-type transition state **3-66ts** to generate a zwitterion intermediate **3-67** with an activation barrier of 47.0 kcal/mol. This insertion is considered to be

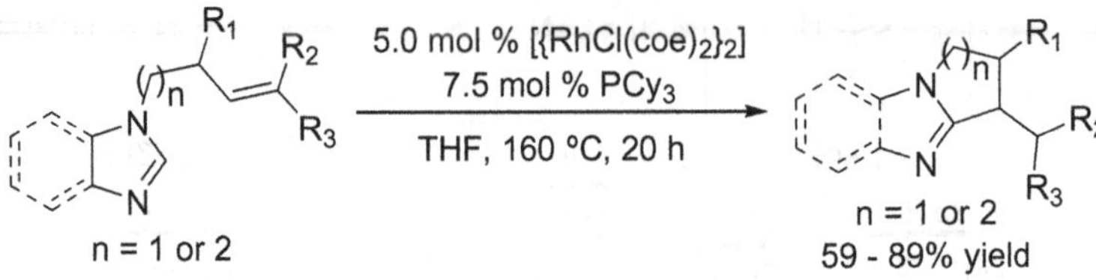

Scheme 3.16 Rh(I)-catalyzed intramolecular C–H alkylation of benzimidazoles

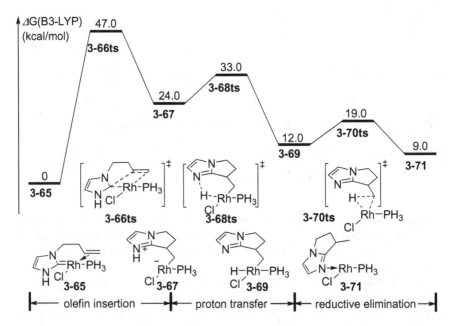

Fig. 3.17 Free energy profiles for the Rh(I)-catalyzed intramolecular C–H alkylation of benzimidazoles. The values are the relative free energies given in kcal/mol calculated at the B3-LYP/LACVP**++//B3-LYP/LACVP** level of theory

the rate-determining step of the catalytic cycle. Then the intramolecular oxidative hydrogen transfer from imidazolium moiety to Rh in **3-67** occurs via transition state **3-68ts** to give the Rh(III)-hydride intermediate **3-69** with an activation barrier of only 9 kcal/mol. Subsequently, a rapid reductive elimination occurs via transition state **3-70ts** to get the product coordinated Rh(I) complex **3-71**. The catalytic cycle is then renewed by the displacement of the product by another molecule of alkene with subsequent C–H activation to form the NHC complex **3-65**.

In 2012, Chang, Jung, and co-workers reported a Rh-catalyzed C–H alkylation of bipyridines with olefins (Scheme 3.18) [49]. In this transformation, the Rh(acac)$_3$ was found to be the most suitable catalyst precursor. When the IMes·HCl was employed, the reaction efficiency and selectivity were significantly increased. A wide range of simple terminal olefins were used as alkylating reagents to give the bisalkylated products. The kinetic isotope effect experiments ($k_{H/D} = 1.01$) suggested that the C–H bond cleavage is excluded in the rate-determining step in the catalytic cycle.

The same group also performed DFT calculations to study the mechanism of this C–H alkylation reaction. The calculated free energy profile for the first alkylation cycle is shown in Fig. 3.19 The catalytic cycle starts with the chelated Rh(I)-NHC complex **3-72**. The first step for the catalytic cycle is the NHC ligand assisted rollover of the bipyridine intermediate **3-72** due to the strong *trans*-effect, which weakens the

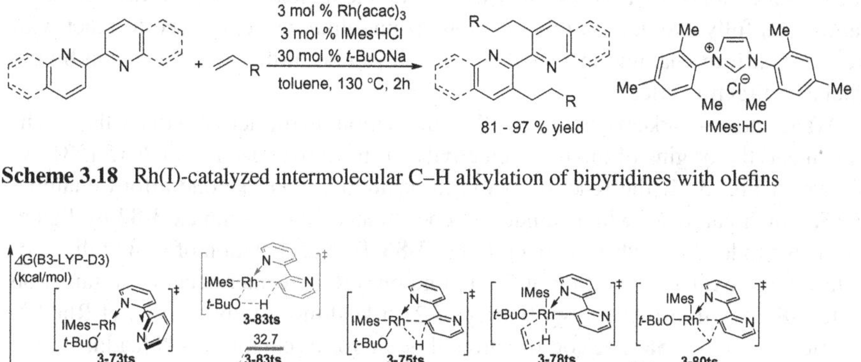

Scheme 3.18 Rh(I)-catalyzed intermolecular C–H alkylation of bipyridines with olefins

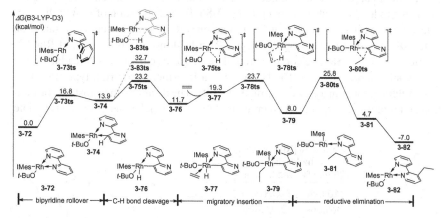

Fig. 3.19 Free energy profiles for the Rh(I)-catalyzed intramolecular C–H alkylation of 2,2′-bipyridines with ethylene. The values are the relative free energies given in kcal/mol calculated at the B3-LYP/6-31G(d)/SRSCECP//B3-LYP-D3/6-31G(d)/SRSCECP level of theory

metal-pyridyl bond positioned *trans* to NHC via transition state **3-73ts** with an activation barrier of 16.8 kcal/mol to give σ-complex **3-74**. The C–H bond activation occurs through an oxidative addition type mechanism via transition state **3-75ts** to generate the Rh-hydride species **3-76**. The σ-CAM type C–H bond activation transition state was also considered. However, the activation free energy of σ-CAM via **3-83ts** is 9.5 kcal/mol higher than that of oxidative addition via **3-75ts**, which indicated that the oxidative addition-type C–H bond activation is favorable. Subsequent coordination of olefin substrate leading to insertion into Rh–H bond occurs relatively easily via transition state **3-78ts** with an activation barrier of 12.0 kcal/mol to afford alkyl Rh(III) intermediate **3-79**. The reductive elimination of complex **3-79** takes place via transition state to generate the alkylated bipyridine coordinated Rh(I) intermediate **3-81**, which would conduct recomplexation to deliver the Rh(I) species **3-82**. The generated species **3-82** would continue decomplexation, rollover cyclometalation, olefin insertion, and reductive elimination to complete the second alkylation.

In 2014, Dong and co-workers [50] developed a ketone α-alkylation reaction using metal-organic cooperative catalysis (MOCC) [51, 52] strategy comprising a secondary amine and Rh(I) complex. As shown in Scheme 3.20, by using [Rh(coe)$_2$Cl]$_2$ (coe = cyclooctene) with NHC ligand and 7-azaindoline as co-catalyst, the α-alkylation of ketones with simple olefins via C(sp^3)-H bond activation can be achieved with high yields. It is worth noting that the reaction only afforded the five-site monoalkylation product with complete regioselectivity, while overalkylation was fully avoided. In this reaction, azaindoline was employed to react with ketone to afford enamine intermediate, which can be further functionalized by the chelated directing effect.

Wang and co-workers performed DFT calculations to further investigate the mechanism and the origins of the regioselectivity of the above reaction in 2015 [53]. As shown in Fig. 3.21, the reaction between azaindoline and ketone afford enamine **3-85,** which can react with azaindoline-coordinated Rh(I) complex **3-84** by ligand exchange to load the substrate in complex **3-86**. By the chelation of azaindoline co-catalyst, the oxidative addition of C–H bond onto Rh(I) center occurs via transition state **3-87ts** with a free energy barrier of 21.5 kcal/mol to give the vinyl Rh(III)-hydride complex **3-88**. The coordination of olefin forms complex **3-89** leading to the sequential insertion into Rh(III)-H bond via a four-membered ring-type transition state **3-90ts** to form an agostic intermediate **3-91**. The disabling the agostic coordination and rotating the equatorial enamine moiety to the axial plane gives intermediate

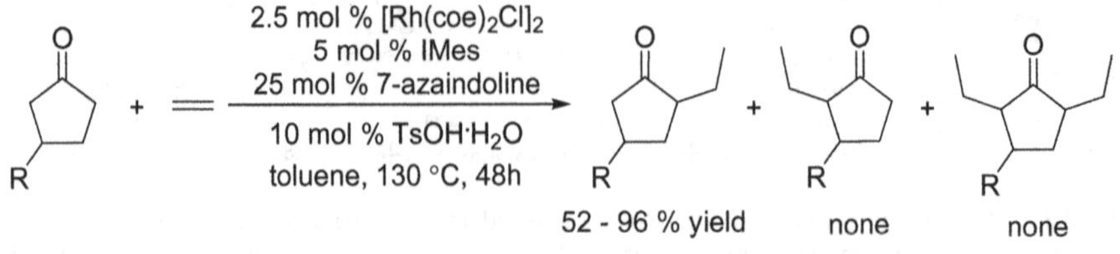

Scheme 3.20 Rh(I)-catalyzed intermolecular C–H alkylation of ketone with olefins

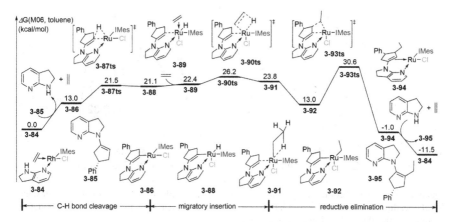

Fig. 3.21 Free energy profiles for the Rh(I)-catalyzed intramolecular C–H alkylation of C–H alkylation of ketone with olefins. The values are the relative free energies given in kcal/mol calculated at the B3-LYP/6-31G(d,p)/SDD//M06/6-31G(d,p)/SDD level of theory in toluene

3-92 with a vacant site *cis* to IMes with exergonic by 10.8 kcal/mol. The subsequent reductive elimination occurs via transition state **3-93ts** with an overall activation energy barrier of 30.6 kcal/mol to afford the alkylated enamine coordinated Rh(I) complex **3-94**. The calculated results show that the reductive elimination is the rate-determining step in the catalytic cycle. The release of the alkylation enamine **3-95** by the ligand exchange with 7-azaindoline and ethylene regenerates the active Rh(I) catalyst **3-84** to complete the catalytic cycle. The final hydrolysis of the alkylated enamine **3-95** can yield α-alkylated ketone product and regenerate 7-azaindoline co-catalyst.

In order to understand the regioselectivity of this reaction, the complete pathways for the alkylation of ketone C2–H bond were also considered. The calculated results also indicated that the rate-determining step is the final reductive elimination via transition state **3-96ts**. The activation energy for the alkylation of ketone C2–H bond via **3-96ts** is 6.1 kcal/mol higher than the corresponding process of C5–H via **3-93ts**, which is well accounting for the experimental complete regioselectivity. The geometry information of transition state **3-96ts** clearly revealed that the bulky IMes ligand restricted the rotation of β-phenyl group in substrate, which leads to steric repulsion between this phenyl group and reacting ethyl group. Therefore, the overalkylation is forbidden by this steric effect (Fig. 3.22).

In 2017, Ellman and co-workers reported a Rh(I)-catalyzed C–H bond *ortho*-alkylation of inactivated azines with α,β-unsaturated carboxylic acid derivatives without an *ortho*-directing group [37]. This transformation has excellent functional group tolerance. Interestingly, the linear or branched product could be generated solely depending on the selection of base (KOPiv or K_3PO_4). In addition, the reaction temperature is 160 °C, which would indicate a high activation free energy for the rate-determining step in the catalytic cycle (Scheme 3.23).

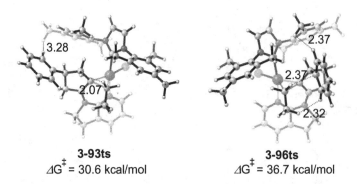

3-93ts   **3-96ts**
$\Delta G^{\ddagger}$ = 30.6 kcal/mol $\Delta G^{\ddagger}$ = 36.7 kcal/mol

Fig. 3.22 Optimized geometries and energy barriers for the reductive elimination transition states **3-93ts** (leading to five-site monoalkylation product) and **3-96ts** (leading to two-site monoalkylation product), with selected bond distances given in Å

Scheme 3.23 Rh(I)-catalyzed C-H activation and *ortho*-alkylation of unactivated azines with acrylamide

To account for the selectivity of the above reaction, Bi and co-workers performed DFT calculation to investigate the detailed mechanism and the role of the base. The computational results clearly reveal that the outer-sphere CMD pathway by using KOPiv as the base is more favorable than the OA-type pathway. As shown in Fig. 3.24, when reacting pyridine coordinates onto Rh(I) in complex **3-97**, a coming KOPiv molecule can assist C-H bond cleavage via transition state **3-98ts** with a free energy barrier of 23.2 kcal/mol to afford a pyridyl Rh(I) intermediate **3-99**. However, the corresponding oxidative addition would bear an energy barrier of 32.0 kcal/mol via transition state **3-108ts**. Then ligand exchange of acrylamide releasing HOPiv and KCl through transition state **3-100ts** generates π-complex **3-109**. The subsequent migratory insertion of acrylamide into C(aryl)-Rh bond generates the alkyl Rh(I) intermediate **3-103** via transition state **3-102ts** with an overall activation free energy of 34.7 kcal/mol, which was considered to be the rate-determining step in the catalytic cycle. The final protonation of intermediate **3-104** via transition state **3-105ts** yields the linear alkylation product **3-106** by the following ligand exchange with new substrate.

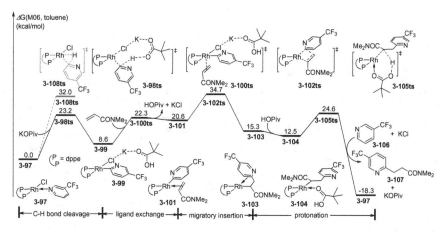

Fig. 3.24 Free energy profiles for the Rh(I)-catalyzed C-H activation and *ortho*-alkylation of unactivated azines with acrylamide using KOPiv as base. The values are the relative free energies given in kcal/mol calculated at the M06/6-31G(d,p)/SDD//B3-LYP/6-31G(d,p)/LANL2DZ level of theory in toluene

When K_3PO_4 was used as the base in this reaction, the OA-type C–H bond activation is preferred to the CMD-type one. As shown in Fig. 3.25, in the participant of K_3PO_4, the OA-type C–H bond activation occurs via transition state **3-110ts** with an activation barrier of 27.3 kcal/mol to afford a Rh(III)-hydride intermediate **3-111**. The following migratory insertion of acrylamide into Rh–H bond occurs through transition state **3-112ts** by overcoming the barrier of 12.1 kcal/mol to afford alkyl-Rh(III) intermediate **3-113**. Then the C(aryl)-C(alkyl) reductive elimination, which is the rate-determining step for the catalytic cycle, occurs via transition state **3-114ts** with an overall activation energy barrier of 30.7 kcal/mol. The branched product **3-115** would be afforded through this process. Alternatively, the outer-sphere CMD type C–H bond activation mechanism is more favorable than the OA-type one. However, the relative free energy of following migratory insertion of acrylamide into C(aryl)-Rh bond transition state **3-119ts** is 13.0 kcal/mol higher than that of **3-114ts**, which indicated that the CMD type C–H bond activation pathway is unfavorable.

3.2.2 C–H Bond Alkylation by Using Alkynes

Alkynes can also be used as alkyl source in Rh-catalyzed C–H bond alkylations through more complicated transformations. In 2014, Li and Chang [42, 54] independently reported a Cp*Rh(III)-catalyzed C–H bond alkylations of quinoline *N*-oxide with alkynes. In these transformations, the *N*-oxide moiety was served as directing group to lead to the regioselective C8–H bond activation. Notably, the *N*-oxide moiety also acts as an endogenous oxidant to keep redox neutral. Furthermore, the O-atom

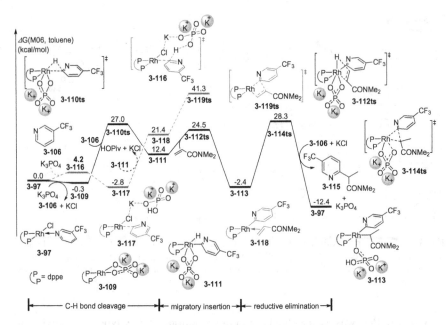

Fig. 3.25 Free energy profiles for the Rh(I)-catalyzed C-H activation and *ortho*-alkylation of unactivated azines with acrylamide using K_3PO_4 as base. The values are the relative free energies given in kcal/mol calculated at the M06/6-31G(d,p)/SDD//B3-LYP/6-31G(d,p)/LANL2DZ level of theory in toluene

transfer also plays a particularly important role to complete the formal alkylation reaction (Scheme 3.26).

The free energy profiles for the preferred pathway considering by DFT calculation for this reaction are given in Fig. 3.27 [55]. The cationic Cp*Rh(III) acetate complex **3-120** was set to the relative zero value for the reaction pathway. The *N*-oxide-directed CMD type C–H bond cleavage takes place via a six-membered ring-type transition state **3-123ts** with an activation free energy of 23.3 kcal/mol resulting in aryl-Rh intermediate **3-124**. The subsequent insertion of the coordinated acetylene into Rh–C(aryl) bond occurs via transition state **3-126ts** with an energy barrier of 18.1 kcal/mol to generate rhodacyclic **3-127**. The following C–O bond reductive

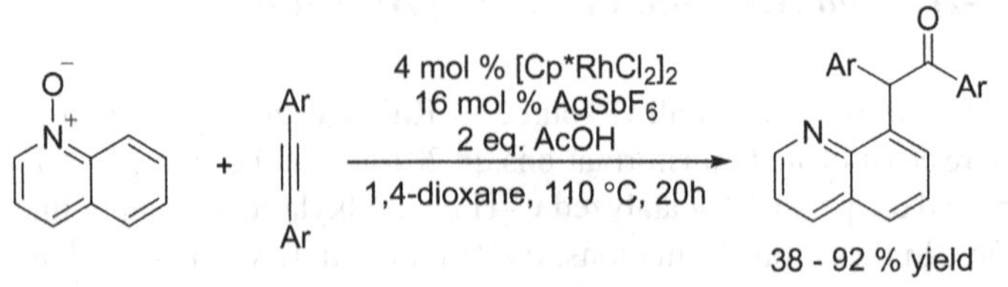

Scheme 3.26 Rh(III)-catalyzed C–H activation and *ortho*-alkylation of quinoline *N*-oxide with alkynes

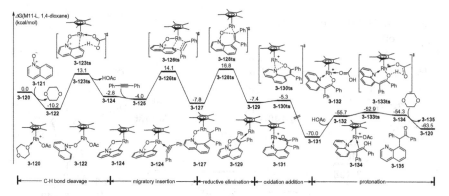

Fig. 3.27 Free energy profiles for the Rh(III)-catalyzed C–H activation and *ortho*-alkylation of quinoline N-oxide with alkynes. The values are the relative free energies given in kcal/mol calculated at the B3-LYP/6-311G(d)/LANL2DZ//M11-L/6-311+G(d)/SDD level of theory in 1,4-dioxane

elimination, which is considered as rate-determining step, leads to the formation of the oxazinoquinolinium-coordinated Rh(I) complex **3-129**. The overall activation energy for this process is 27.0 kcal/mol. Then, the oxidative addition of N–O bond onto Rh(I) takes place via transition state **3-130ts** to give π-enolate Rh complex **3-131** with 64.7 kcal/mol exothermic. The final protonation of intermediate **3-131** gives the enol complex, which would undertake isomerization to provide the alkylation product **3-135** and regenerate active catalyst **3-120**.

In another example, Chang and co-workers applied a similar strategy to accomplish the selective Rh(III)-catalyzed coupling reaction of arylnitrones with internal alkynes [8]. This cyclization gives rise to indolines in good yields with moderate to high diastereoselectivity. The reaction that proceeded in the presence of excessive $H_2^{18}O$ indicated that the O-atom transfer (OAT) occurs via an intramolecular manner. The kinetic isotope effect ($k_H/k_D = 1.92$) shows that the C–H bond cleavage may be involved in the rate-limiting step (Scheme 3.28).

Lan [56] and Chen [57] independently performed theoretical calculations to investigate the mechanism, regio- and diastereoselectivity for this cyclization reaction. The calculated free energy profiles for the imine insertion and protonation steps are shown in Fig. 3.29 When the π-enolate Rh(III) complex **3-139** was generated, the imine

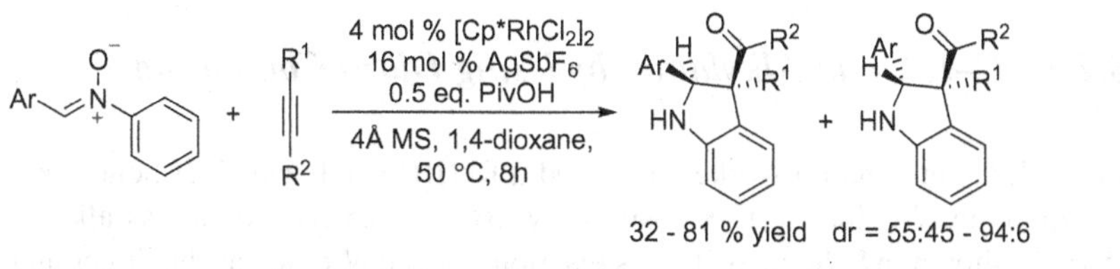

Scheme 3.28 Rh(III)-catalyzed C–H activation and cyclization of arylnitrones with internal alkynes

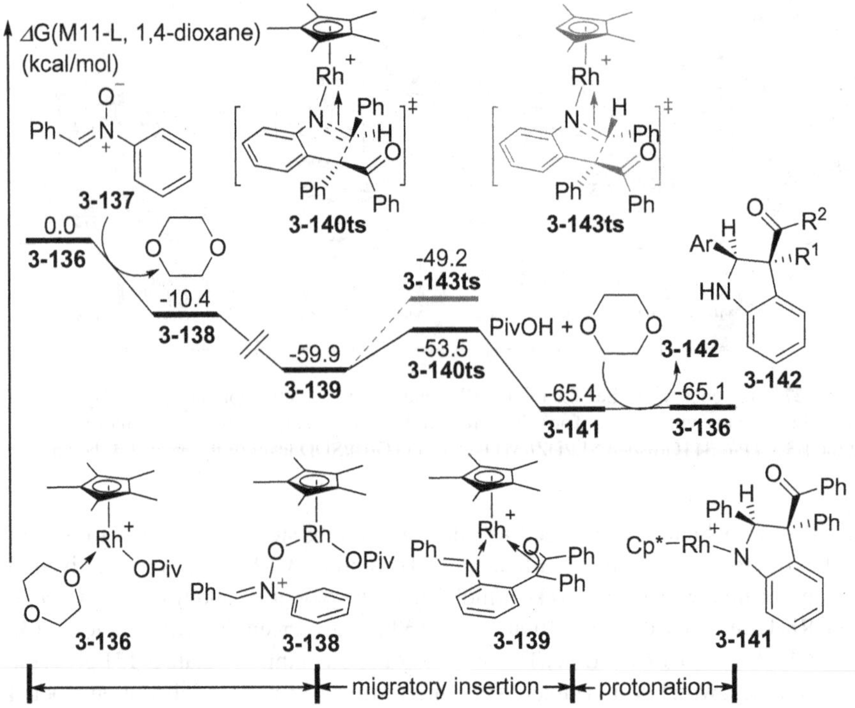

Fig. 3.29 Free energy profiles for the Rh(III)-catalyzed C–H activation and cyclization of arylni-trones with internal alkynes. The values are the relative free energies given in kcal/mol calculated at the M11-L/6-311+G(d)/LANL08(f)//B3-LYP/6-311G(d)/LANL08(f) level of theory in 1,4-dioxane

would insert into C(benzyl)-Rh bond via transition state **3-140ts** to form *threo*-amino-Rh(III) complex **3-141** with a free energy barrier of 6.4 kcal/mol. The complex **3-141** could be protonated by pivalic acid to yield the corresponding *threo*-type product **3-142**. Alternatively, the imine insertion also could occur via transition state **3-143ts**, which would lead to formation of *erythto*-type product. The strong repulsion of the two phenyl groups on the forming C–C bond in **3-143ts** attributes to its high activation free energy. The calculated results are consistent with the experimental observations.

3.2.3 C–H Bond Alkylation by Using Diazo Compound

In 2014, Chang and co-workers reported a Rh(III)-catalyzed C–H bond alkylation of quinoline *N*-oxides at C-8 position by using diazo compounds as alkyl source [58]. As shown in Scheme 3.30, this reaction utilized *N*-oxide as the directing group to assist the remote C–H functionalization of quinolones, which proceeded highly

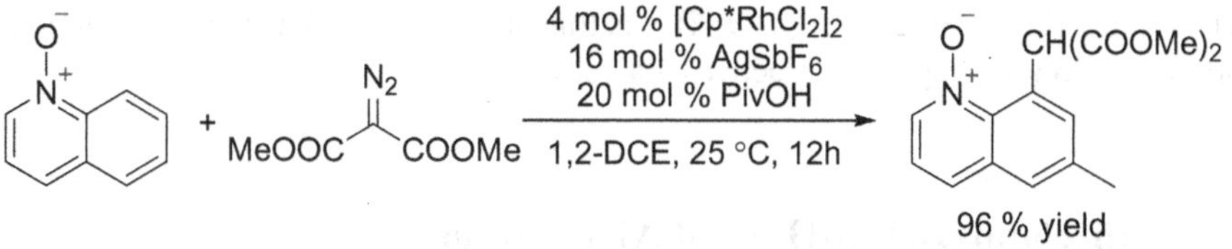

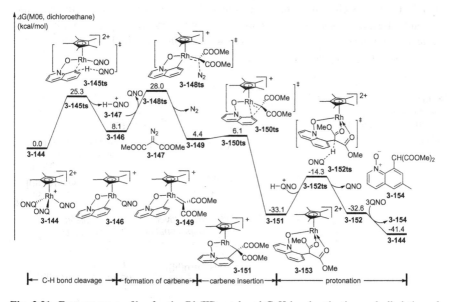

Scheme 3.30 Rh(III)-catalyzed C–H bond activation and alkylation of quinoline N-oxides at C-8 position

efficiently at room temperature with excellent regioselectivity and functional group tolerance.

Lin and co-workers [59] carry out DFT calculations to investigate the mechanism of this reaction. As shown in Fig. 3.31, the catalytic cycle starts from a cationic Cp*Rh(III) complex **3-144**. An outer-sphere electrophilic deprotonation by extra quinoline N-oxide takes place via transition state **3-145ts** with an activation barrier of 25.3 kcal/mol to generate five-membered rhodacycle intermediate **3-146**. The coordination of diazo compound onto Rh(III) leads to carbenation via transition state **3-148ts** with an overall energy barrier of 28.0 kcal/mol. Consequently, it was considered to be rate-determining step. The subsequent migratory insertion of carbene into the C(aryl)–Rh bond occurs via transition state **3-150ts** with a barrier of only 1.7 kcal/mol to generate the six-membered rhodacycle intermediate **3-151**.

Fig. 3.31 Free energy profiles for the Rh(III)-catalyzed C–H bond activation and alkylation of quinoline N-oxides at C-8 position. The values are the relative free energies given in kcal/mol calculated at the M06/6-311++G(d,p)/LANL2DZ//B3-LYP/6-31G(d)/LANL2DZ level of theory in dichloroethane

The protonation of **3-151** by [QNO-H]$^+$ via transition state **3-152ts** yields alkylation product **3-154** and regenerates active catalyst **3-144**.

3.3   Rh-Catalyzed C–H Bond Alkenylation

3.3.1   C–H Bond Alkenylation by Using Olefins

Olefins can be considered as the nucleophile to undertake oxidative coupling with the C–H bond in arenes using a Rh(III) catalyst in the presence of an external oxidant, which can be mechanistically considered as an oxidative Heck-type coupling [13, 60–75]. As shown in Scheme 3.32, the C–H bond cleavage usually occurs through a CMD-type mechanism to form aryl-Rh(III) intermediate **3-156**. The subsequent migratory insertion of olefin into C–Rh bond generates the alkyl-Rh(III) complex **3-157**. Then β-H elimination results alkenylation product and the hydride-Rh(III) complex **3-158**, which can be reductively deprotonated by extra base. The generated Rh(I) species **3-160** can be oxidized by the external oxidant to regenerate the active Rh(III) catalyst **3-155**.

In 2011, Liu and co-workers [76] reported a Rh(III)-catalyzed oxidative Heck-type coupling reaction of protected phenols with olefins (Scheme 3.33). In this reaction, electron-deficient olefins, such as acrylates, can be used as the alkenyl source, where [Cp*RhCl$_2$]$_2$ was chosen as the catalyst and Cu(OAc)$_2$ was chosen as the external oxidant. The kinetic isotope effect ($k_H/k_D = 3.1$) was observed experimentally, which indicated that the C–H bond cleavage is most likely involved in the rate-determining step in the catalytic cycle.

Scheme 3.32 General mechanism for the Rh(III)-catalyzed C–H bond alkenylation by using olefins

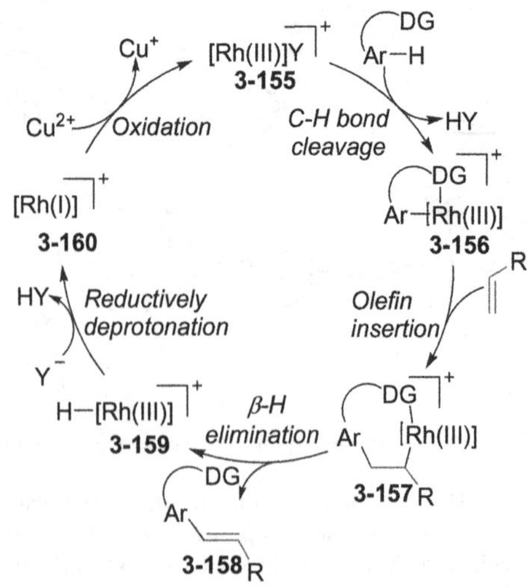

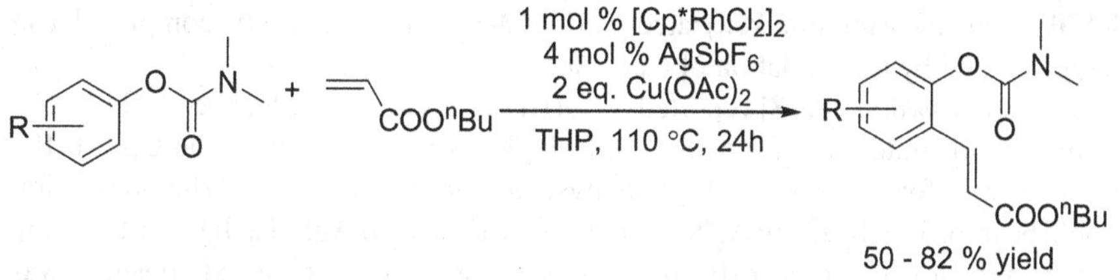

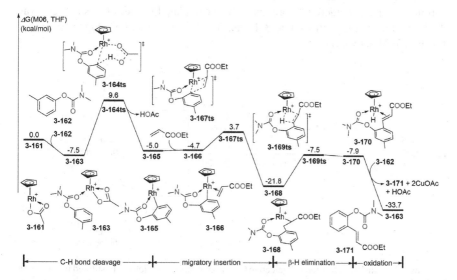

Scheme 3.33 Rh(III)-catalyzed oxidative Heck coupling reaction of protected phenols with olefin

Fu and co-workers [77] carried out a DFT study to reveal the detailed mechanism of this reaction. As shown in Fig. 3.34, the cationic complex $[CpRh(OAc)]^+$ **3-161** was selected as relative zero in calculated free energy profiles. The coordination of carbamate **3-162** onto Rh(III) center gives intermediate **3-163**. The C–H bond activation occurs through a CMD-type mechanism via transition state **3-164ts** with a free energy barrier of 17.1 kcal/mol to form an aryl Rh(III) intermediate **3-165**. The calculated results indicated that the CMD-type C–H bond cleavage is the rate-determining step in the catalytic cycle, which is consistent with the experimental observation. The coordination of acrylates to the Rh(III) center in **3-165** gives intermediate with 0.3 kcal/mol endothermic. The subsequent migratory insertion of C = C double bond into the Rh–C(aryl) bond occurs via transition state **3-167ts** to form alkyl Rh(III) intermediate **3-168**, which undergoes β-H elimination through four-membered ring transition state **3-169ts** to afford a Rh(III)-hydride intermediate

Fig. 3.34 Free energy profiles for Rh(III)-catalyzed oxidative Heck coupling reaction of *m*-tolyl dimethylcarbamate with ethyl acrylate. The values are the relative free energies given in kcal/mol calculated at the M06/6-311+G(d,p)/LANL2DZ//B3-LYP/6-31G(d)/LANL2DZ level of theory in THF

3-170. After releasing of alkenylated product **3-171**, active catalytic complex **3-163** is regenerated by the oxidation of $Cu(OAc)_2$.

Ma and co-workers [78] reported a Rh(III)-catalyzed C7-selective C–H alkenylation of substituted indoles using N-pivaloyl as directing group. The Cp*Rh(III) complex was found to be the best catalyst in this transformation. The conversion would be improved by using $AgNTf_2$ as additive instead of $AgSbF_6$. It is worth noting that this reaction would majorly afford 2-alkenylated indole if the sixth position was substituted by CF_3 group (Scheme 3.35).

Liu and co-workers [79] performed DFT calculations to investigate the mechanism, regioselectivity, and substituent effect of the above reaction. The calculated free energy profiles for the C7-alkenylation of N-acyl indole are shown in Fig. 3.36. The reaction starts with the coordination of the N-acyl indole onto the Rh(III) in intermediate **3-174**. The NTf_2-assisted C7-selective C–H bond activation occurs via transition state **3-175ts** with a free energy barrier of 23.5 kcal/mol to generate aryl-Rh(III) intermediate **3-176**. The subsequent migratory insertion of the $C = C$ double bond in the coordinated acrylate into the C(aryl)–Rh bond occurs via transition state **3-178ts** to afford alkyl-Rh(III) intermediate **3-179** with a barrier of 13.0 kcal/mol. The alkyl-Rh(III) **3-179** could isomerize to an agnostic intermediate **3-180** leading to the following β-H elimination via transition state **3-181ts** with an energy barrier of 8.9 kcal/mol to give a hydride Rh(III) complex **3-182**. Finally, the active catalyst **3-172** is regenerated by the release of 7-olefination product **3-182** and the oxidation of $Cu(OAc)_2$. The calculated results reveal that the rate-determining step for the catalytic cycle is the CMD type C–H bond cleavage. The overall activation free energy for the catalytic cycle is 23.5 kcal/mol. As a contrast, the C2–H bond activation was also considered theoretically. The calculated barrier of this process was 26.5 kcal/mol via transition state **3-184ts**, which is 17.6 kcal/mol higher than that for the C7–H activation.

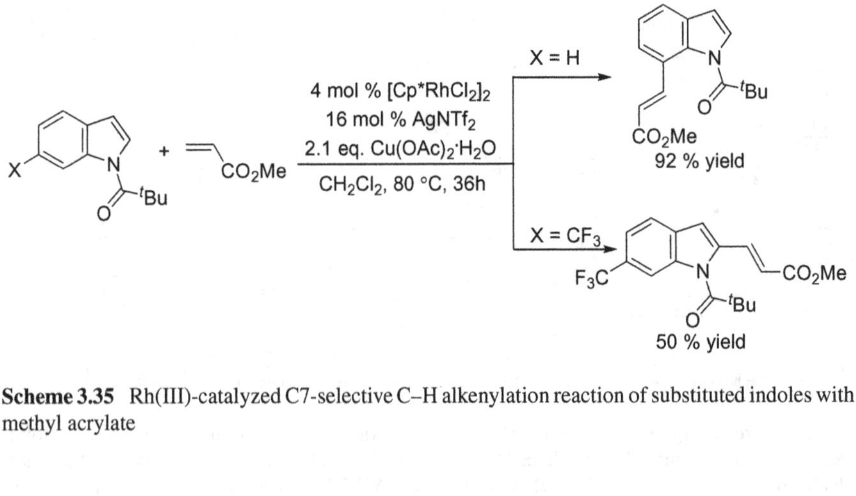

Scheme 3.35 Rh(III)-catalyzed C7-selective C–H alkenylation reaction of substituted indoles with methyl acrylate

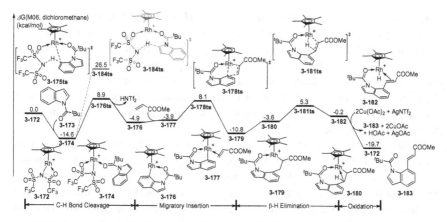

Fig. 3.36 Free energy profiles for Rh(III)-catalyzed C7-selective C–H alkenylation reaction of *N*-acyl indole with methyl acrylate. The values are the relative free energies given in kcal/mol calculated at the M06/6-311++G(d,p)/LANL2DZ//B3-LYP/6-31G(d,p)/LANL2DZ level of theory in dichloromethane

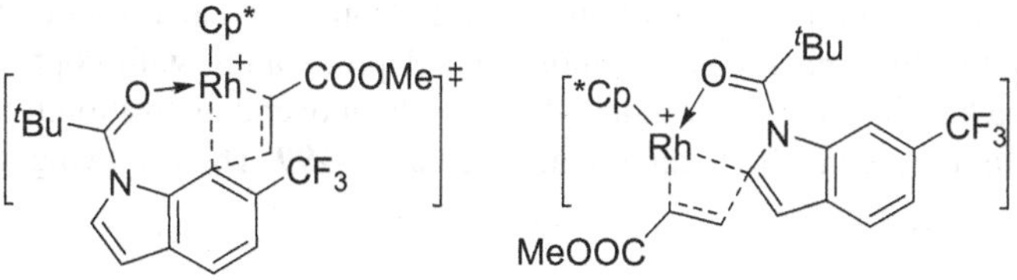

3-185ts

$\Delta G^{\ne}$ = 29.7 kcal/mol

3-186ts

$\Delta G^{\ne}$ = 23.9 kcal/mol

Fig. 3.37 The activation free energies for the olefin insertion transition states **3-185ts** and **3-186ts**

The substituent effect of C–H bond activation and alkenylation has also been considered theoretically. When 6-trifluoromethyl *N*-acyl indole is used, the rate-determining step is changed to the olefin insertion step. As shown in Fig. 3.37, the computational results indicated that the activation free energy for the olefin insertion via transition state **3-185ts** leading to C7–H bond alkenylation is 5.8 kcal/mol higher than that via **3-186ts** leading to C2–H bond alkenylation. The repulsion between acrylate and CF$_3$ group in **3-185ts** contributes to the difference of activation free energy. The computational results are in good agreement with the experimental observations.

3.3.2 C–H Bond Alkenylation by Using Alkynes

The alkynes can be used as coupling partners in Rh-catalyzed C–H bond alkenylation reactions [80–96]. This transformation can be considered to be a redox-neutral process of formal alkyne insertion into an aryl C–H bond. The non-redox mechanism can be summarized in Scheme 3.38a, where Rh(III)-catalyzed C–H bond activation starts with a CMD type C(aryl)–H bond cleavage. The following alkyne insertion and protonation would result in the final alkenylation product. In an alternative redox-involved mechanism (Scheme 3.38b), the oxidative addition of C(aryl)–H bond onto Rh(I) species provide an aryl hydride-Rh(III) intermediate. After acetylene insertion into Rh–H bond, C(aryl)-C(vinyl) reductive elimination also could yield alkenylation product and revive Rh(I) species.

In 2014, Wang and co-workers [96] reported a Rh(III)-catalyzed alkenylation reactions of 8-methylquinolines with alkynes undergoing a C(sp^3)–H bond activation. The KIE was found to be $k_H/k_D = 4.0$ indicating that the C–H bond cleavage was involved in the rate-determining step. The deuterium-labeling experiments revealed that the C(sp^3)–H bond activation is a reversible process. The deuterium-labeling experiments also indicated that protonolysis may be involved in the catalytic cycle (Scheme 3.39).

In 2015, Morokuma and co-workers [97] performed a detailed theoretical study to investigate the mechanism of the above reaction. As shown in Fig. 3.40, the coordination of 8-methylquinoline in complex **3-196** leads to the activation of benzylic C–H bond, which can form a hydrogen bond with the coming copper acetate complex. The C–H activation undergoes intermolecular deprotonation with copper acetate complex as a base via transition state **3-197ts** with an overall activation free energy of 26.4 kcal/mol to give alkyl-Rh(III) intermediate **3-199**. The following acetylene

Scheme 3.38 General mechanism for the Rh-catalyzed C–H bond alkenylation by using alkynes

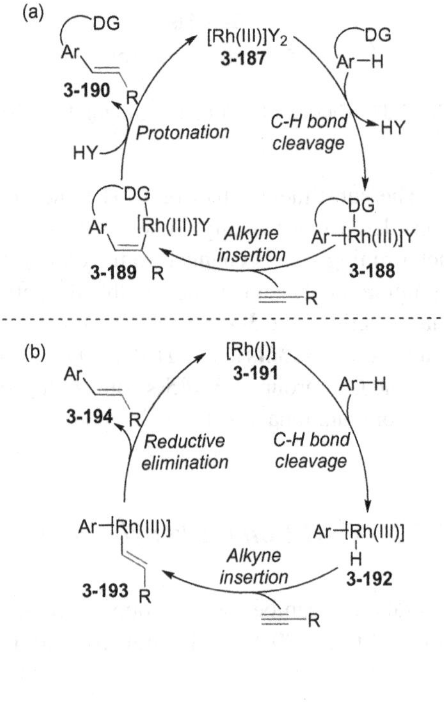

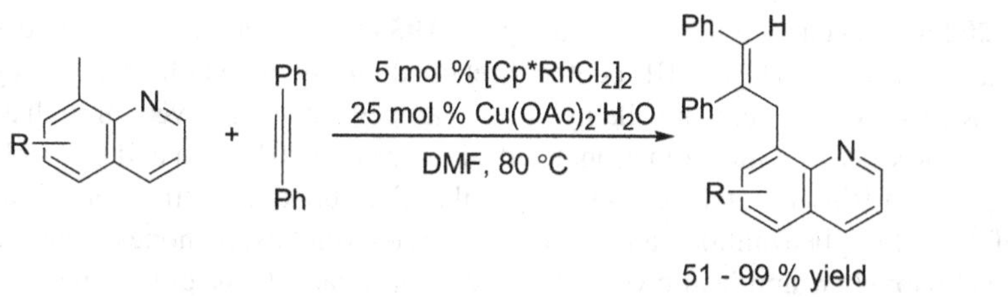

Scheme 3.39 Rh(III)-catalyzed alkenylation reactions of 8-methylquinolines with alkynes by C(sp^3)–H bond activation

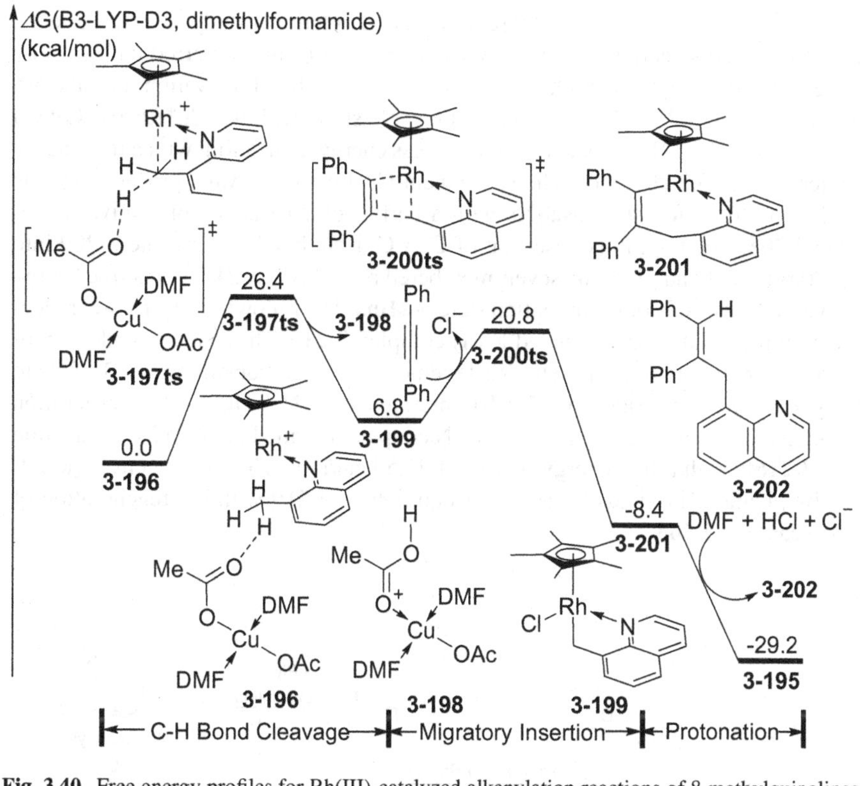

Fig. 3.40 Free energy profiles for Rh(III)-catalyzed alkenylation reactions of 8-methylquinolines with alkynes by C(sp^3)–H bond activation. The values are the relative free energies given in kcal/mol calculated at the B3-LYP/6-311G(d,p)/SDD//B3-LYP/6-311G(d,p)/SDD level of theory in dimethylformamide

insertion takes place via transition state **3-200ts** with an activation energy barrier of 19.2 kcal/mol to give the alkenyl-Rh(III) complex **3-201** after dissociation of chloride. The final protonation and ligand exchange yield the alkenylation product

3-202 and regenerate the active catalyst **3-195** to complete the catalytic cycle. The oxidation state of the Rh(III) center is retained during the whole catalytic cycle.

Rh-catalyzed C–H bond alkenylation of arenes can be achieved through a complicated process with multifunctionalized substrates. In 2014, Lin and co-workers [98] reported a Rh(III)-catalyzed intermolecular C–H bond activation and alkenylation of *N*-hydroxybenzamides and alkyne-tethered cyclohexadienones. The use of *N*-pivaloyloxybenzamide allows the formation of tetracyclic isoquinolones through an *N*-Michael addition process. When acyl is used instead of pivaloyl, the formation of hydrobenzofurans was allowed through a *C*-Michael addition process. In addition, the KIE experiments indicated that the C–H bond activation was involved in the rate-determining step (Scheme 3.41).

In 2016, Li and co-workers [99] reported a computational study to investigate the mechanism and selectivity. The free energy profiles for the preferred pathways by using *N*-pivaloyloxybenzamide in Rh-catalyzed C–H bond activation and alkenylation are given in Fig. 3.42. The catalytic cycle starts with the Cp*Rh(III)(OPiv)$_2$ active catalyst. The N–H bond activation is less energy demanding with a free energy barrier of only 3.2 kcal/mol. The subsequent C–H bond cleavage proceeds via the CMD type transition state **3-207ts** with a barrier of 20.3 kcal/mol to give rhodacycle **3-208**. The migratory insertion of C $\equiv$ C triple bond in substrate **3-209** into the C(aryl)-Rh bond gives the seven-membered rhodacycle **3-211** with a free energy barrier of 15.9 kcal/mol via transition state **3-210ts**. The following C(vinyl)-N reductive elimination affords a chelated Rh(I) complex **3-213**. The oxidation addition of the N-O bond in organic part onto Rh(I) center occurs via transition state **3-214ts** to exergonically generate an amido Rh(III) intermediate **3-215**. The *N*-Michael addition of **3-215** then takes place to afford the C–N coupled intermediate **3-217** via transition state **3-216ts** with a free energy barrier of 21.5 kcal/mol. The subsequent protonolysis by pivalic acid yields tetracyclic isoquinolone **3-219** with the regeneration of active catalyst **3-203**.

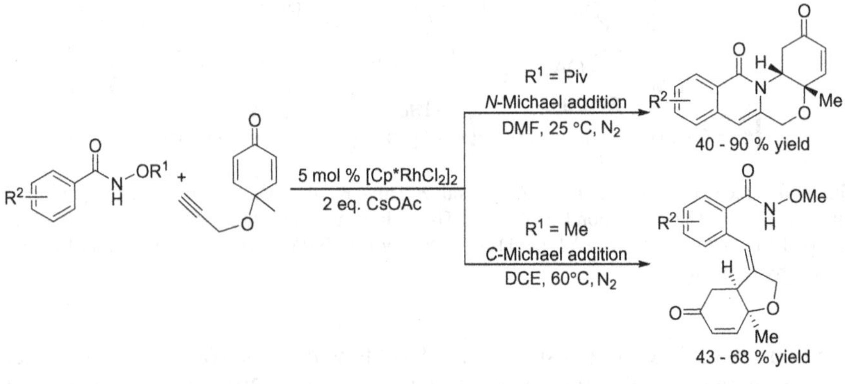

Scheme 3.41 Rh(III)-catalyzed intermolecular C–H bond activation and alkenylation of O-substituted *N*-hydroxybenzamides and alkyne-tethered cyclohexadienone

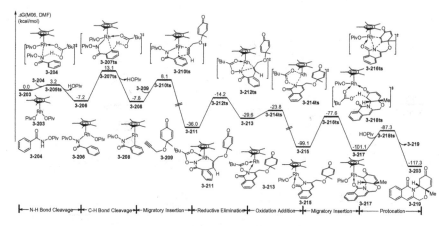

Fig. 3.42 Free energy profiles for Rh(III)-catalyzed intramolecular C–H bond activation and alkenylation of O-Piv N-hydroxybenzamides and alkyne-tethered cyclohexadienone. The values are the relative free energies given in kcal/mol calculated at the M06/6-311+G(d,p)/SDD//M06/6-31G(d)/LANL2DZ level of theory in DMF

When N-acetoxybenzamide is used as an alternative substrate, seven-membered rhodacycle **3-220** can be formed through a similar process. However, a *C*-Michael addition was detected theoretically as the favorable pathway to afford alkyl Rh(III) intermediate **3-222** via transition state **3-221ts** with an activation energy barrier of 17.2 kcal/mol. The sequential deprotonation yields alkenylation product **3-226** and regenerates the active catalyst **3-203** (Fig. 3.43).

The internal oxidation by directing group can be used to keep redox neutral in Rh-catalyzed C–H bond alkenylation by using acetylene, where bifunctionalization of acetylenes could be achieved. In 2013, Liu and Lu et al. [96] reported a Rh(III)-catalyzed C–H alkenylation reaction of *N*-phenoxyacetamides with alkynes. The N–O bond in *N*-phenoxyacetamides acts as the internal oxidant to keep redox neutral. In this transformation, a subtle change of solvent (e.g. methanol or CH$_2$Cl$_2$) can result in the formation of either *ortho*-hydroxyphenyl enamides or benzofurans with high chemoselectivity (Scheme 3.44).

In 2016, Wu and Houk et al. [100] reported a theoretical study to investigate the mechanism of this reaction. As shown in Fig. 3.45, the Cp*Rh(OAc)$_2$ is set to relative zero in the calculated free energy profiles. The consecutive N–H and C–H activations were considered to undergo CMD process to afford a five-membered rhodacycle **3-232**, via transition states **3-229ts** and **3-231ts**, respectively. The subsequent alkyne insertion into C(aryl)-Rh bond occurs via transition state **3-234ts** with a free energy barrier of 25.4 kcal/mol to give the ring-extended intermediate **3-235**. The following phenoxy migration from N to Rh occurs via a three-membered ring-type transition state **3-236ts** leading to the cleavage of N–O bond with an energy barrier of 24.6 kcal/mol to form a Rh–nitrene complex **3-237**. Then nitrene moiety inserts

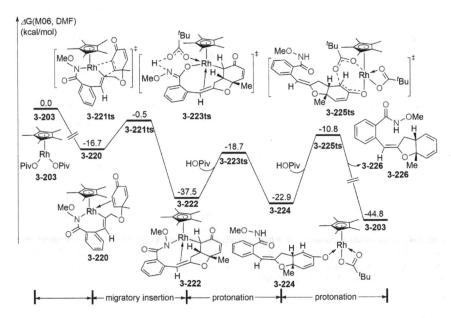

Fig. 3.43 Free energy profiles for Rh(III)-catalyzed intramolecular C–H bond activation and alkenylation of O-Me N-hydroxybenzamides and alkyne-tethered cyclohexadienone. The values are the relative free energies given in kcal/mol calculated at the M06/6-311+G(d,p)/SDD//M06/6-31G(d)/LANL2DZ level of theory in DMF

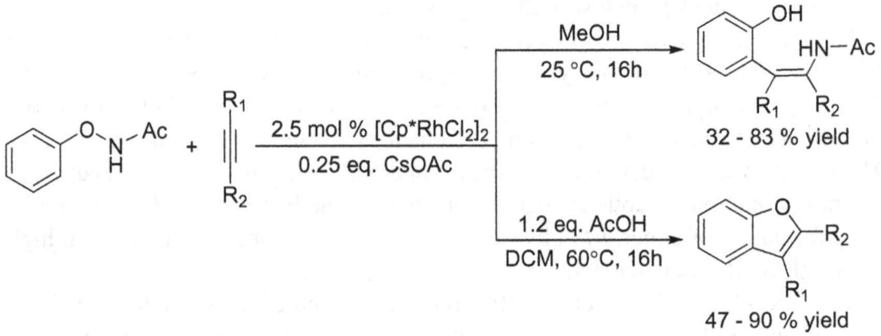

Scheme 3.44 Rhodium(III)-catalyzed C–H alkenylation reaction of N-phenoxyacetamides with alkynes

into C(aryl)–Rh bond via transition state **3-238ts** to afford an amido Rh(III) intermediate **3-239**. Two sequential protonations yield *ortho*-hydroxyphenyl enamide **3-240** (Fig. 3.45).

As shown in Fig. 3.46, when acetate acid was used as an additive, the protonation of Rh–nitrene complex **3-237** provides a hydrogen bond complex, which restricts the insertion of nitrene. Alternatively, a C(vinyl)-O(phenoxy) reductive elimination

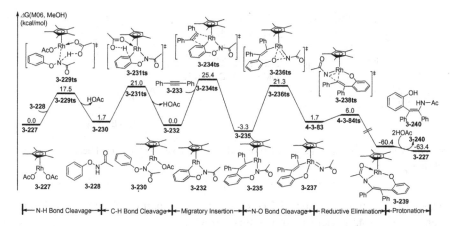

Fig. 3.45 Free energy profiles for Rhodium(III)-catalyzed C–H alkenylation reaction of *N*-phenoxyacetamides with alkynes. The values are the relative free energies given in kcal/mol calculated at the M06/6-311++G(d,p)/SDD//B3-LYP/6-31G(d)/LANL2DZ level of theory in MeOH

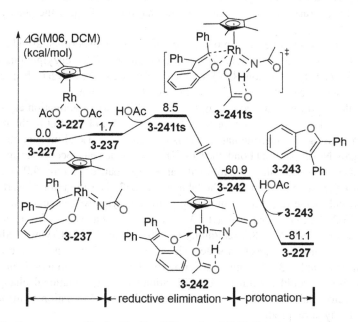

Fig. 3.46 Free energy profiles for Rhodium(III)-catalyzed C–H alkenylation and cyclization reaction of *N*-phenoxyacetamides with alkynes. The values are the relative free energies given in kcal/mol calculated at the B3-LYP/6-31G(d)/LANL2DZ//M06/6-311++G(d,p)/SDD level of theory in DCM

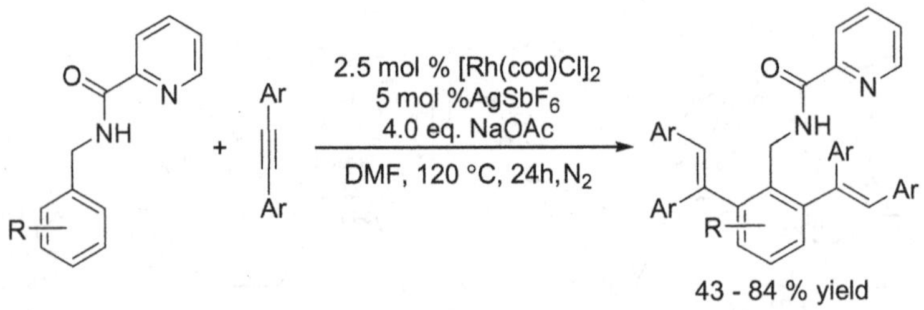

Scheme 3.47   Rh(I)-catalyzed C-H activation and *ortho*-olefination of *N*-benzylpicolinamide with alkynes

takes place via transition state **3-241ts** with an energy barrier of only 6.8 kcal/mol. Benzofuran **3-243** is yielded through this pathway.

Oxidative addition is an alternative of the C–H bond activation, when low-valence Rh(I) species was involved in arene alkenylations, which can provide hydride aryl Rh(III) intermediate. The migratory insertion of alkyne reactant into Rh(III)-H bond followed by a C(aryl)-C(vinyl) reductive elimination would give the alkenylation product.

Carretero and co-workers [84] reported a combination of experimental and computational study of Rh(I)-catalyzed *ortho*-di-olefination of *N*-benzylpicolinamides (Scheme 3.47). In this reaction, Rh(I) complex was used as the catalyst in the presence of NaOAc, which can offer a good yield of di-olefinated arenes.

The calculated rational free energy profiles for this *ortho*-olefination are shown in Fig. 3.48. The substrate coordinated Rh(I) species **3-244** was set to the relative zero point. The agostic intermediate **3-245** is formed with 3.6 kcal/mol endothermic. Directing by N–Rh covalent bond, the C(aryl)–H bond cleavage occurs via an oxidation addition transition state **3-246ts** with an energy barrier of only 9.0 kcal/mol, which provides a five-membered rhodabicycle **3-247**. The migratory insertion of alkyne, which is the rate-determining step in the catalytic cycle, occurs via transition state **3-249ts** to form vinyl-Rh(III) **3-250**. The overall activation free energy for this migratory insertion is 18.2 kcal/mol. The subsequent C(aryl)-C(vinyl) reductive elimination occurs via transition state **3-251ts** to irreversibly generate the mono-alkenylation complex **3-252** with a barrier of 16.7 kcal/mol. The mono-alkenylation complex **3-252** would continue decomplexation and conformational changes to achieve cyclometalation, alkyne insertion, and reductive elimination to afford the final di-alkenylation product.

3.4   Rh-Catalyzed C–H Bond Alkynylation

In the presence of Rh-catalyst, C(aryl)–H considered as nucleophile can react with electrophilic alkynyl to achieve direct arene alkylation in a cross-coupling of C(aryl)-C(alkynyl) [101–107]. In this reaction, ethynyl high-valent iodoxolone was chosen as

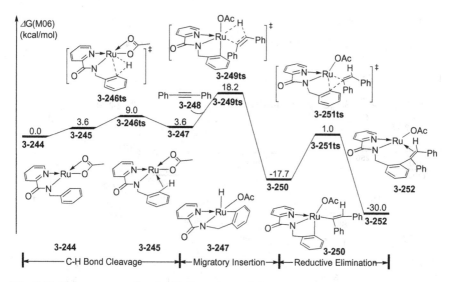

Fig. 3.48 Free energy profiles for Rh(I)-catalyzed C–H activation and *ortho*-olefination of *N*-benzylpicolinamide with alkynes. The values are the relative free energies given in kcal/mol calculated at the M06/6-311+G(d,p)/SDD//M06/6-31G(d)/LANL2DZ level of theory

alkynylation reagent, which was first reported by Waser in the alkynylation of indoles under gold or palladium catalysis [108–112]. The typical alkynylation reagent (1-[(triisopropylsilyl)ethynyl]-1,2-benziodoxol-3(1H)-one (TIPS-EBX)) was successfully applied in Rh-catalyzed arene alkynylations by Li [113–115], Loh [116–118], and Glorius [119], independently. In 2016, Patil and co-workers [120] reported a Rh(III)-catalyzed site-selective C-8 alkynylation of isoquinolones. In this reaction, [Cp*RhCl₂]₂ was used as catalyst in the presence of AgSbF₆ additive. A broad range of synthetically useful functional groups (-F, -Cl, -Br, -CF₃, -OMe, alkyl, etc.) were tolerated in this transformation (Scheme 3.49).

In 2018, Liu and co-workers [121] performed theoretical studies to reveal the intrinsic mechanism of this reaction. The calculated free energy profiles of this reaction are shown in Fig. 3.50. Active species **3-254** can be formed by decomposition of [Cp*RhCl₂]₂ and AgSbF₆. The C8–H bond activation of isoquinolone then proceeds via transition state **3-257ts**, where the O center of TIPS-EBX **3-253** acts as a Brønsted

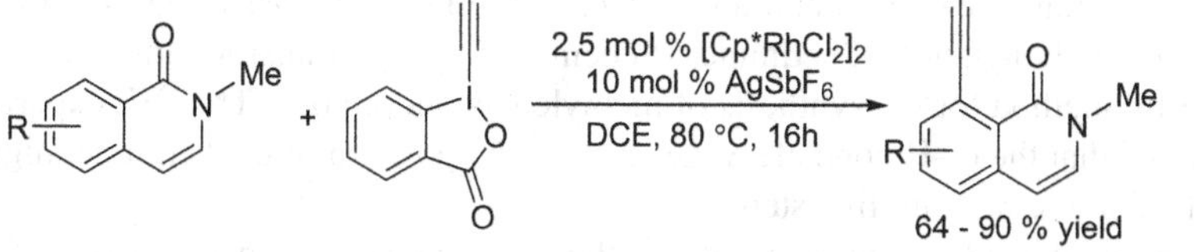

Scheme 3.49 Rh(III)-catalyzed C-8 alkynylation reaction of isoquinolones with TIPS-EBX

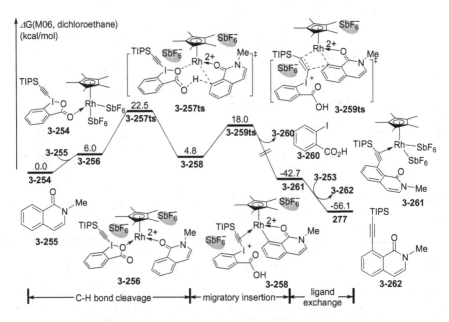

Fig. 3.50 Free energy profiles for Rh(III)-catalyzed C-8 alkynylation reaction of isoquinolones with TIPS-EBX. The values are the relative free energies given in kcal/mol calculated at the M06/6-311++G(d,p)/SDD//B3-LYP/6-31G(d,p)/LANL2DZ level of theory in dichloroethane

base to activate the C8–H bond in **3-255** with a barrier of 22.5 kcal/mol to give the five-membered rhodacycle **3-258**. In this step, the cleavage of I-O bond occurs simultaneously. The subsequent Rh-mediated nucleophilic substitution of ethynyliodonium with aryl takes place via transition state **3-259ts** forming C(acetyl)-C(aryl) bond by the release of iodobenzene **3-260**. Active species **3-254** can be regenerated by the coordination of TIPS-EBX **3-253** and the release of alkynylation product **3-262**.

3.5   Rh-Catalyzed C–H Bond Carbonylation

In Rh-catalyzed C–H functionalization reactions, the C–H bond cleavage leads to the formation of a C-Rh bond, which can undergo a following migratory insertion of CO to realize carbonylation [122–127]. In 2016, Huang and Lan [128] reported a combination of experimental and theoretical studies of Rh(III)-catalyzed oxidative cyclocarbonylation of ketimines (Scheme 3.51). This transformation provided a powerful method for the synthesis of methyleneisoindolinone. The KIE experiment indicated that the C–H bond cleavage at the *ortho* position of the ketimine might be involved in the rate-limiting step.

The calculated free energy profile for the dominant reaction pathway of cyclocarbonylation is shown in Fig. 3.52, which starts with the active Rh(III) catalyst **3-263**.

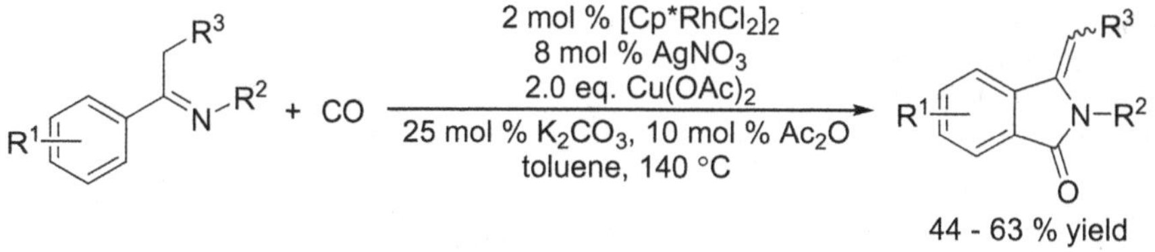

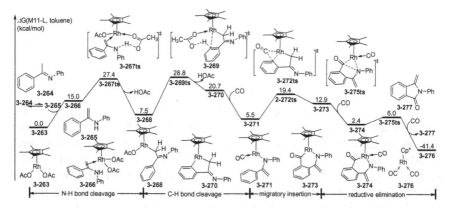

Scheme 3.51 Rh(III)-catalyzed oxidative cyclocarbonylation reaction of ketimines

Fig. 3.52 Free energy profiles for Rh(III)-catalyzed C–H bond activation and oxidative cyclo-carbonylation reaction of ketimines. The values are the relative free energies given in kcal/mol calculated at the M11-L/6-311+G(d)/SDD//B3-LYP/6-31G(d)/SDD level of theory in toluene

The active catalyst **3-263** was coordinated by enamine **3-265** to form π-complex **3-266**. The cleavage of the N–H bond occurs via transition state **3-267ts** leading to the iminoethyl-Rh(III) **3-268** with a free energy barrier of 27.4 kcal/mol. The following C(aryl)–H bond activation proceeds via transition state **3-269ts** with a free energy barrier of 21.3 kcal/mol to form a five-membered rhodacycle **3-270**. The sequential imine-enamine tautomerization and CO coordination form the intermediate **3-271**. The migratory insertion of CO into Rh–C(aryl) bond then takes place to give a six-membered rhodacycle **3-273** via transition state **3-272ts**. Rhodacycle **3-273** can be stabilized by the further coordination of another molecule of carbon monoxide to afford **3-274**. A rapid reductive elimination yields methyleneisoindolinone product **3-277** and provides a CO-coordinated Rh(I) complex **3-276**, which can be oxidized by Cu(OAc)₂ to give an active Rh(III) catalyst **3-263** to complete the catalytic cycle.

3.6 Rh-Catalyzed C–H Bond Annulation

Rh-catalyzed aromatic C–H bonds activation assisted by a directing group represents an increasingly important strategy for the construction of new covalent bonds,

Scheme 3.53 The
classification of directing
groups in Rh-catalyzed C–H
bond annulations

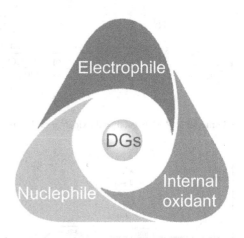

because it inherently enables the efficient construction of organic building blocks to prepare a wide range of functional compounds [2, 17, 129–131]. However, the directing group itself in most such reactions remains intact during the reaction and would be removed after the C–H functionalization process, which leads to the lower atom and step economy. One possible way to circumvent this drawback is developing certain effective C–H annulation processes, in which the directing group could be utilized as a reactive moiety to be incorporated into the desired product [132–135]. Through this strategy, Rh-catalyzed C–H bond activation and subsequent cyclization with alkynes and alkenes have evolved into one of the most powerful methods in the construction of various heterocycles [60, 136–139]. Mechanistically, directing group plays an important role, which can be utilized as a directing group to participate in the following transformation in Rh-catalyzed C–H bond activation and annulation reactions (Scheme 3.53).

3.6.1   Directing Group as Electrophile

The unsaturated groups often can play as π-acid ligand, which would be employed as directing group in Rh-mediated C–H activation [20, 140–143]. After C–H activation, the unsaturated bond can undergo a nucleophilic attack to further functionalize in annulations. In this type of reaction, both redox and non-redox mechanisms are reasonable depending on the C–H activation step. As shown in Scheme 3.54a, the redox mechanism involves oxidative addition of C(aryl)-H bond onto Rh(I) species, migratory insertion of the unsaturated directing group, and C(aryl)-C(alkyl) reductive elimination. In redox-involved mechanism (Scheme 3.54b), the Rh(III)-catalyzed C–H bond activation occurs via a CMD type mechanism. The following olefin insertion, unsaturated directing group insertion, and protonation would give the annulation product.

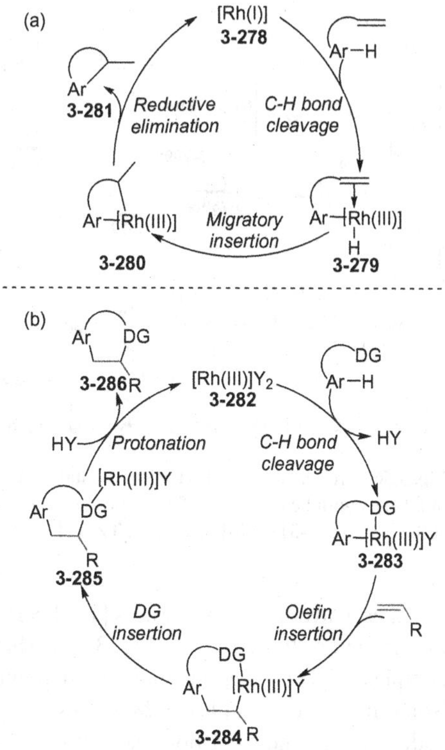

Scheme 3.54 General mechanism for Rh-catalyzed C–H bond annulation with directing group as electrophile

The alkenyl groups can act as weak π-acid ligands to direct C–H bond activation and functional substrate in the Rh-catalyzed C–H bond activation, which can undergo a further nucleophilic addition to achieve annulation [144–146]. In 2001, Bergman, Ellman, and co-workers [48] developed a Rh(I)-catalyzed C–H bond activation and intramolecular coupling of alkenyl heterocycles to create heterobicycles. In this transformation, alkenyl group can be used as electrophilic directing group to realize C–H bond activation and functionalization of imidazole, which can be used to construct both five- and six-membered rings (Scheme 3.55).

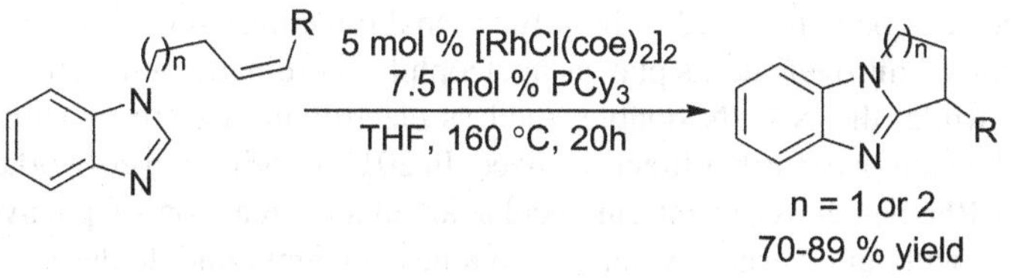

n = 1 or 2
70-89 % yield

Scheme 3.55 Rh(I)-catalyzed C–H bond activation and intramolecular coupling of alkenes with heterocycles to create two-functionalized azoles

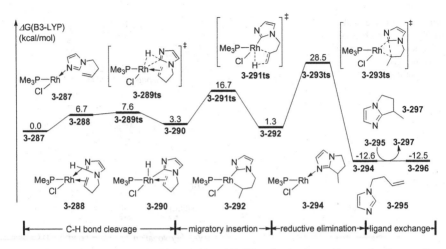

Fig. 3.56 Free energy profiles for Rh(I)-catalyzed C–H bond activation and intramolecular coupling of alkenes with heterocycles. The values are the relative free energies given in kcal/mol calculated at the B3-LYP/6-311+G(d,p)/LANL2TZ+(3f)//B3-LYP/6-31G(d)/LANL2DZ level of theory

Yates, Cavell, and co-workers [147] performed a DFT study to reveal the detailed mechanism of this reaction (Fig. 3.56). The simplified imidazole coordinated Rh(I) complex **3-287** is set as the relative zero point in the calculated free energy profile. The isomerization of complex **3-287** gives an agostic structure **3-288** with 6.7 kcal/mol endothermic. The C–H bond activation takes place through an oxidation addition-type mechanism via transition state **3-289ts** with a free energy barrier of only 7.6 kcal/mol to give the aryl hydride-Rh(III) intermediate **3-290**. The alkenyl group using as directing group can electrophilically react with hydride in a migratory insertion via transition state to generate a six-membered rhodacycle **3-292**. The activation free energy for the insertion is small at 13.4 kcal/mol. The C(aryl)-C(alkyl) reductive elimination of intermediate **3-292** gives the annulation product-coordinated Rh(I) intermediate **3-294**. Standing at 27.2 kcal/mol, the energy barrier required for the reductive elimination transition state **3-293ts** is the highest individual barrier along the reaction pathway. In the final step of the cycle, the coordination of imidazole substrate **3-295** and release carbocyclization product **3-297** regenerate the active catalyst **3-296**.

Diazenes can be considered to be bifunctional directing group, where the lone-pair electron on nitrogen atoms play as nucleophile coordinating onto Rh to achieve chelation, while, the N = N double bond as electrophile can react with coming nucleophilic pattern to further functionalized. In 2015, Glorius and co-workers [148] reported a Rh(III)-catalyzed intermolecular annulation reaction of phenyldiazene and alkenes without external oxidants to synthesis N-aminoindole derivatives. The Cp*Rh(III) complex was used as the active catalyst in this transformation. A variety of electron-deficient alkenes and conjugative dienes could be efficiently converted into the corresponding products (Scheme 3.57).

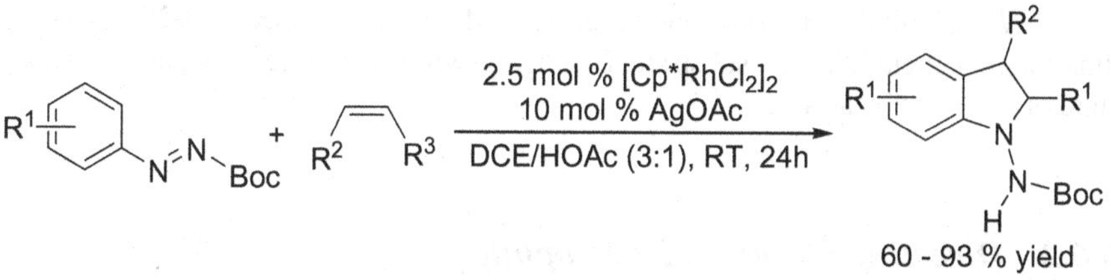

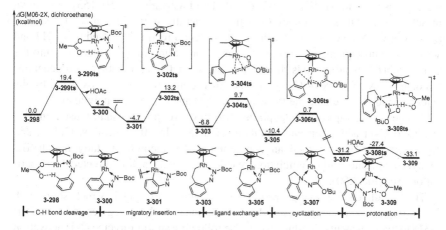

Scheme 3.57 Rh(III)-catalyzed intermolecular annulation reaction of aryl-substituted diazenecarboxylates with alkenes

Fig. 3.58 Free energy profiles for Rh(III)-catalyzed intermolecular annulation reaction of aryl-substituted diazenecarboxylates with alkenes. The values are the relative free energies given in kcal/mol calculated at the M06-2X/6-311+G(d,p)/LANL2DZ//M06-2X/6-31G(d)/LANL2DZ level of theory in dichloroethane

As shown in Fig. 3.58, DFT calculations were performed by Huang and coworkers [149] to reveal the mechanism of the above Rh(III)-catalyzed intermolecular annulation reaction. The calculated free energy profiles start from a phenyldiazene substrate coordinated Rh(III) species **3-298**, which can undergo a acetate-assisted CMD type C–H bond activation via transition state **3-299ts** with a free energy barrier of 19.4 kcal/mol to generate the five-membered rhodacycle **3-300**. The migratory insertion of the coordinated alkene into Rh–C(aryl) bond proceeds via transition state **3-302ts** with an activation barrier of 17.9 kcal/mol to exergonically give seven-membered rhodacycle **3-303**. The following interesting coordination switch between two nitrogen atoms in diazene forms six-membered isomer **3-305**. The further coordination of N-Boc group increases the electrophilicity of N = N double bond, which can react with alkyl group on Rh through a formal 1,4-insertion via transition state **3-306ts**. The generated amino Rh(III) intermediate **3-307** can undergo protonolysis to release *N*-aminoindole product and regenerate Rh(III) active species **3-298** by coordination of a phenyldiazene substrate. During the whole catalytic cycle, the oxidative state of Rh remains at +3.

3.6.2 *Directing Group as Nucleophile*

In the presence of an external oxidant, the oxidative coupling of C(aryl)-H with nucleophilic directing groups can give annulation product in the presence of extra unsaturated substrates in Rh(III)-catalyzed C–H bond activations [150–153]. As shown in Scheme 3.59, these annulation reactions generally start with Rh-catalyzed X-H (X = N or O) bond cleavage to achieve a covalent direction followed by C–H bond activation. Subsequently, the migratory insertion of unsaturated substrates into the C(aryl)-Rh(III) bond extends rhodacycle. The reductive elimination then generates the other C–X (X = N or O) bond. The Rh(I) complex would be oxidized by the external oxidant to regenerate the active catalyst and complete catalytic cycle.

Amides as a typical nucleophilic directing group are usually used in Rh-catalyzed C–H activation and annulation reaction. In 2013, Moisés, Gulías, and co-workers [154] performed an experimental and theoretical study of Rh(III)-catalyzed intramolecular annulation reaction involving amide-directed C–H activation (Scheme 3.60).

The calculated free energy profiles are shown in Fig. 3.61, which starts with RhCp*(OAc)$_2$. The acetate-assisted deprotonation of N–H bond gives amido Rh(III) **3-318** to load covalent directing group. The following amido-directed C–H activation occurs via CMD-type transition state **3-319ts** to give the aryl-Rh(III) intermediate **3-320** with an energy barrier of 24.3 kcal/mol. The subsequent migratory insertion of the coordinated alkyne moiety into the C(aryl)-Rh bond, which is usually invoked in the intermolecular cases, occurs via transition state **3-321ts** to afford seven-membered rhodacycle **3-322**. The following reductive elimination takes place via **3-323ts** to

Scheme 3.59 General mechanism for the Rh(III)-catalyzed C–H bond activation and annulation reactions using nucleophilic directing group

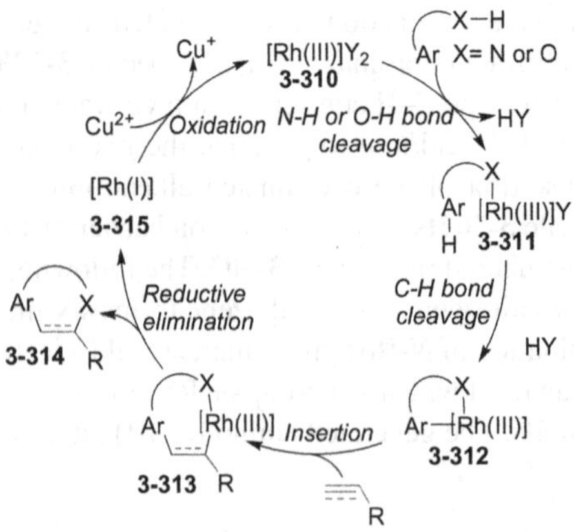

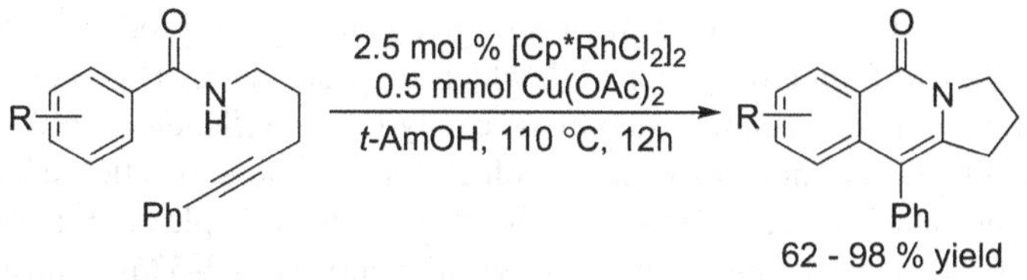

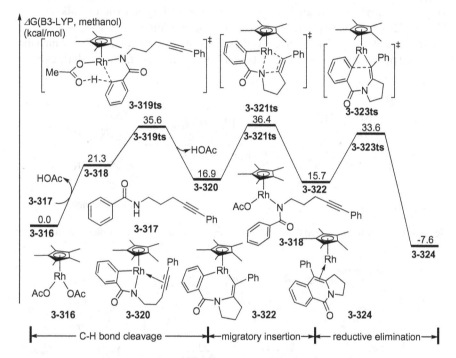

Scheme 3.60 Rh(III)-catalyzed intramolecular C–H activation and annulation reaction of alkyne tethered benzamides

Fig. 3.61 Free energy profiles for Rh(III)-catalyzed intramolecular C–H activation and annulation reaction of alkyne tethered benzamides. The values are the relative free energies given in kcal/mol calculated at the B3-LYP/6-311+G(d,p)/SDD//B3-LYP/6-31G(d)/LANL2DZ level of theory in methanol

give the alkenyl-coordinated Rh(I) complex **3-324**. **3-324** would deliver annulation product and a Rh(I) complex, which can be oxidized by Cu(OAc)$_2$ to complete the catalytic cycle. In this reaction, both C(aryl)-H bond and amido-directing group play as nucleophile to react with alkyne moiety, while two redundant electrons are formally removed by exogenous oxidant.

In 2014, Macgregor and co-workers [155] performed experimental and theoretical studies of Rh(III)-catalyzed intermolecular C–H bond annulation of pyrazoles with alkynes. In this reaction, pyrazole acts as a nucleophilic directing group to undergo deprotonation and coupling with alkynes to give heterocyclic products. The k_H/k_D value for KIE was found to be about 2.7, which suggests that C–H bond cleavage is involved in the rate-determining step in the catalytic cycle (Scheme 3.62).

The calculated rational free energy profiles for the above reaction are given in Fig. 3.63 The catalytic cycle starts with a hydrogen bond complex **3-325**, which can undergo a N–H bond activation in pyrazole via transition state **3-326ts** with an energy barrier of 11.5 kcal/mol. The C–H bond activation assisted by acetate ligand occurs via transition state **3-328ts** with an overall activation free energy of 21.4 kcal/mol to afford five-membered rhodacycle **3-329** in which both the nucleophilic C(aryl)-H and pyrazolyl are activated. The subsequent migratory insertion of alkyne into C(aryl)-Rh bond takes place via transition state **3-330ts** to extend rhodacycle in intermediate **3-331**. A rapid reductive elimination gives complex **3-333**, in which the produced pyrazoloisoquinoline is η^4-bound to the Rh(I) center. The catalytic cycle is

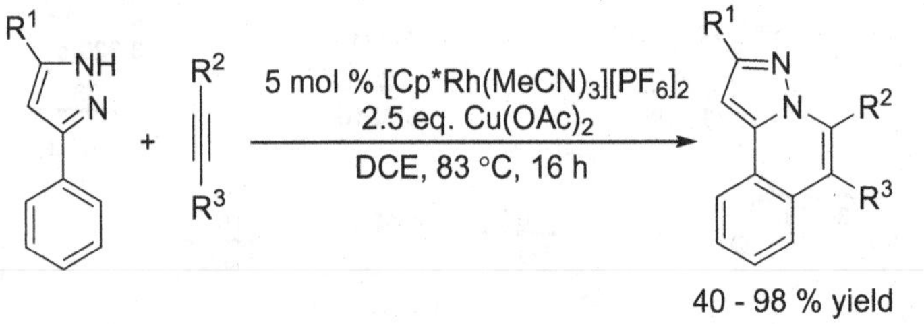

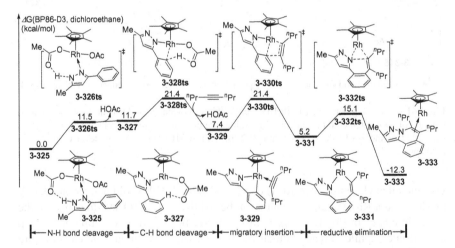

Scheme 3.62 Rh(III)-catalyzed C–H bond activation and annulation of pyrazoles with alkynes

Fig. 3.63 Free energy profiles for Rh(III)-catalyzed C–H bond activation and annulation of pyrazoles with alkynes. The values are the relative free energies given in kcal/mol calculated at the BP86-D3/6-31G(d,p)/RECPs//BP86/6-31G(d,p)/RECPs level of theory in dichloroethane

completed by the release of the annulation product and reoxidation of an unspecified Rh(I) species to the active Rh(III) catalyst.

In another work, You and co-workers [156] reported the first example of the Rh(I)-catalyzed asymmetric dearomatization of phenyl naphthol with C(aryl)–H activation using phenolic hydroxyl as nucleophilic directing group. The chiral cyclopentadienyl (Cp) Rh(III) was used as catalyst to get the chiral naphthalenone products with up to 98% yield and 94% ee under the optimized conditions. The experimental study suggested that the C–H cleavage might be involved in the turnover-limiting step (Scheme 3.64).

They also performed a DFT calculation to probe the mechanistic details of this asymmetric dearomatization reaction of phenylnaphthols (Fig. 3.65) [157]. The acetate assisted deprotonation of the hydroxyl group of phenylnaphthol occurs via transition state **3-335ts** to generate a naphtholate-Rh(III) **3-337**. The released acetic acid was neutralized by K_2CO_3 to give **3-336**. With the *ortho*-C–H bonds to approach the Rh center, the C–H bond cleavage proceeds via a six-membered CMD-type transition state **3-338ts** with an energy barrier of 17.3 kcal/mol. The calculated results also indicated that the C–H bond activation is the rate-determining step in the catalytic cycle. Again, this acetic acid was neutralized by K_2CO_3 to irreversibly deliver the six-membered rhodacyclic **3-339**. Through the migratory insertion via transition state **3-341ts**, rhodacycle is extended in **3-342**. The subsequent dearomatization process can be achieved via a formal reductive elimination transition state **3-343ts** with an energy barrier of 14.8 kcal/mol to generate naphthalenone-coordinated Rh(I) complex **3-344**. The catalytic cycle is finally completed by the releasing of the desired product **3-346** and the regeneration of active Rh(III) species **3-345** by oxidation with $Cu(OAc)_2$. DFT calculations found that the enantioselectivity is controlled by the acetylene insertion, which would undergo four possible transition states as **3-341ts**, **3-347ts**, **3-348ts**, and **3-349ts**. The Steric effect can be observed in other three processes, which leads to the major pathway takes place via transition state **3-341ts**.

The phosphoryl group is one of the most crucial chemical motifs in organic chemistry. A large number of transition metal-catalyzed methods have been developed to construct the structurally sophisticated organophosphorus compounds. The phosphoryl group can be utilized as nucleophilic directing group to functionalize alkynes/olefins constructing the organophosphorus compounds. In 2013, Lee and

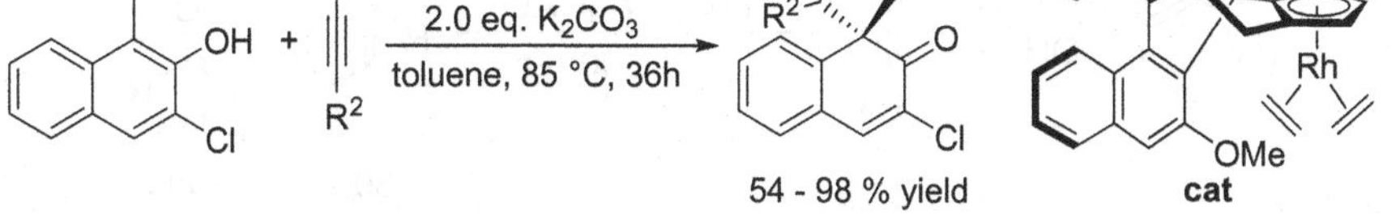

Scheme 3.64 Rh(I)-catalyzed C–H activation and asymmetric dearomatization reaction of 2-naphthols

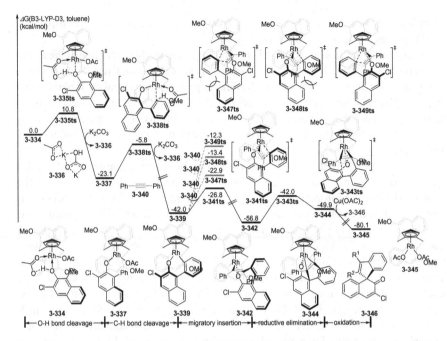

Fig. 3.65 Free energy profiles for Rh(I)-catalyzed C–H activation and asymmetric dearomatization reaction of 2-naphthols. The values are the relative free energies given in kcal/mol calculated at the B3-LYP-D3/6-31G(d,p)/def2-TZVPP//ωB97XD/6-31G(d,p)/SDD level of theory in toluene

co-workers [158] reported a Rh(III)-catalyzed phosphoryl-directed C–H bond activation and annulation of arylphosphonic acid monoesters with alkynes for the synthesis of phosphaisocoumarins with excellent yields and functional group tolerance. The Cp*Rh(III) complex was used as the active catalyst in this transformation. The KIE experiment ($k_H/k_D = 5.3$) indicated that the *ortho* C–H bond cleavage is involved in the rate-determining step (Scheme 3.66).

Zhao group [159] performed a theoretical study to investigate the detailed mechanism of the above reaction in 2014. As shown in Fig. 3.67, the active CpRh(III) catalyst **3-351** was generated by reaction of CpRh(III)ClO$_2$ with AgOAc. The

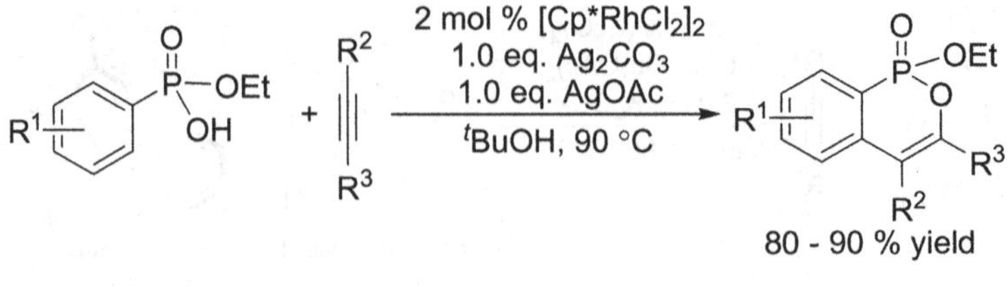

Scheme 3.66 Rh(III)-catalyzed phosphoryl-directed C–H bond activation and annulation of arylphosphonic acid monoesters s with alkyne

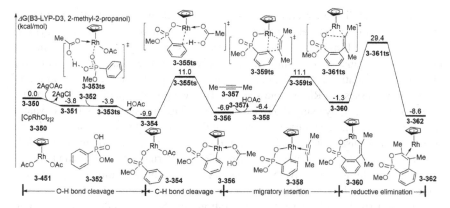

Fig. 3.67 Free energy profiles for Rh(III)-catalyzed phosphoryl-directed C–H bond activation and annulation of alkynes with arylphosphonic acid monoesters. The values are the relative energies given in kcal/mol calculated at the B3-LYP-D3/6-311++G(d,p)/SDD//B3-LYP/6-31G(d)/LANL2DZ level of theory in 2-methyl-2-propanol

hydrogen transfer from phosphonic acid to acetate occurs via transition state **3-353ts** to generate the O–Rh bond intermediate **3-354**. The covalent directed C–H bond cleavage proceeds via a CMD-type transition state **3-355ts** to form the five-membered rhodacycle **3-356** with a free energy barrier of 20.9 kcal/mol. The following migratory insertion of the coordinated alkyne into the C(aryl)-Rh bond takes place via **3-359ts** with a barrier of 17.5 kcal/mol to form extended rhodacycle **3-360**. The C(alkenyl)-O reductive elimination would give the annulation product coordinated Rh(I) complex **3-362**. The Rh(I) would be oxidized by Ag_2CO_3 to Rh(III) to complete the catalytic cycle. The C(alkenyl)-O reductive elimination was considered to be the rate-determining step in the catalytic cycle with an overall activation free energy of 30.7 kcal/mol.

3.6.3 Directing Group as Internal Oxidants

In lots of experimental cases, Cp*Rh(III) can be used as competent catalyst for the C–H bond activation under oxidative conditions, which allows easily accessible starting materials [160–162]. The strategy typically makes use of the directing group to realize cyclometalation at the C–H bond. Then, the insertion of unsaturated bonds can be followed by reductive elimination to yield the desired heterocycle. The Cp*Rh(III)-catalyzed oxidative coupling of arene C–H bonds with olefins, alkynes, or aldehydes generally require external oxidants like copper or silver salts to regenerate the catalyst and complete the catalytic cycle [60, 65, 163, 164]. When directing groups can act as internal oxidants, external oxidant was unnecessary to turn over the high-valence Rh catalyst.

The use of oxidizing directing groups as internal oxidants has recently emerged as an attractive strategy in C–H activation, which majorly involves *N*-oxide, *N*-acyloxy, *N*-methoxy, *N*-sulfinyl, hydrazine, and nitroso groups [165–171]. Two common proposed mechanisms are summarized in Scheme 3.68, which involves reductive elimination–oxidative addition processes via Rh(I) species and group migration–insertion processes via Rh–nitrene species. The two possible pathways start from a N–H bond cleavage by using Rh(III) species **3-363** to load substrate with a covalent directing group. Then a C(aryl)-H activation through CMD process followed by an intermolecular unsaturated bond insertion results in the common rhodacycle **3-366**. In pathway A, reductive elimination forms a Rh(I) intermediate, which can undergo an oxidative addition by active N–X bond (X = O, N, or S) in directing group and sequential protonolysis to release annulation product and regenerate Rh(III) catalyst. Alternatively (pathway B), an X group migration from N to Rh results a Rh(V)-nitrene complex, which can undergo a nitrene insertion to achieve annulation.

In 2011, Guimond and co-workers [172] reported an experimental and theoretical study of Rh(III)-catalyzed redox-neutral C–H bond activation using *N*-pivaloyloxy amide as oxidizing directing groups to synthesize isoquinolone. This reaction tolerates terminal alkynes as well as alkenes to access isoquinolones with various substitution patterns. The kinetic isotope effect ($k_H/k_D = 15$) indicated that the C–H bond cleavage would be involved in the rate-determining step in the catalytic cycle (Scheme 3.69).

The calculated free energy profiles of the dominant reaction pathway through reductive elimination–oxidative addition are shown in Fig. 3.70. The catalytic cycle starts from CpRh(III) complex **3-470**, which undergoes a deprotonation to load covalent directing group in complex **3-472**. A CMD type C(aryl)–H bond activation occurs via transition state **3-473ts** to form a five-membered rhodacycle **3-474** with an overall

Scheme 3.68 General mechanism for the Rh(III)-catalyzed C–H bond activation and annulation reaction using directing group as internal oxidant

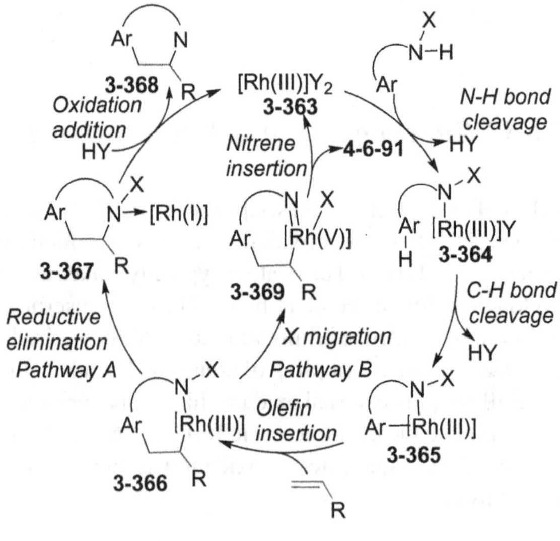

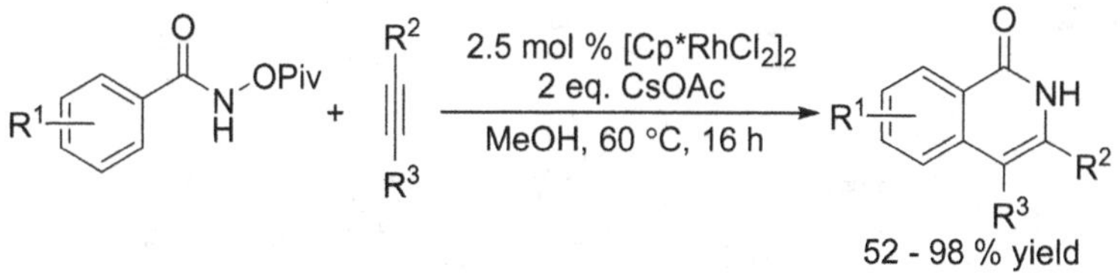

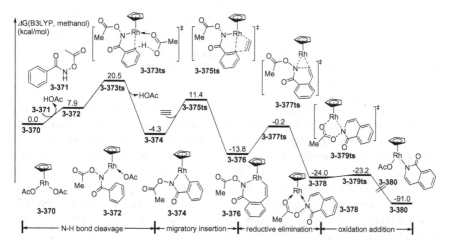

Scheme 3.69 Rh(III)-catalyzed redox-neutral C–H bond activation using *N*-pivaloyloxy amide as oxidizing directing groups to synthesis isoquinolone

Fig. 3.70 Free energy profiles for Rh(III)-catalyzed redox-neutral C–H bond activation using *N*-pivaloyloxy amide as oxidizing directing groups to synthesize isoquinolone. The values are the relative free energies given in kcal/mol calculated at the B3-LYP/TZVP/DZVP//B3-LYP/TZVP/DZVP level of theory in methanol

activation free energy of 20.5 kcal/mol. The intermolecular acetylene insertion via transition state **3-475ts** extends rhodacycle to seven-membered ring in intermediate **3-476**. Subsequently, a C(aryl)-N(amido) reductive elimination achieves annulation and afford Rh(I) in complex **3-478**. The calculated energy barrier of reductive elimination is 13.6 kcal/mol. Then, a rapid oxidative addition leading to N–O bond cleavage takes place via a five-membered ring-type transition state **3-379ts** to result amido-Rh(III) species **3-480**. The Rh(III) active catalyst can be regenerated through further protonolysis by acetic acid.

The analogous Rh(III)-catalyzed (4 + 2) annulation of *N*-pivaloyloxy benzamide with ethylene has also been studied computationally by Xia and co-workers [173], where the alternative migration–insertion pathway was considered to be favorable. As shown in Fig. 3.71, starting from Cp*Rh(III) complex **3-381**, N–H bond cleavage followed by CMD type C(aryl)-H activation, five-membered rhodacycle **3-387** is given, which can undergo an intermolecular ethylene insertion to afford Rh(III)

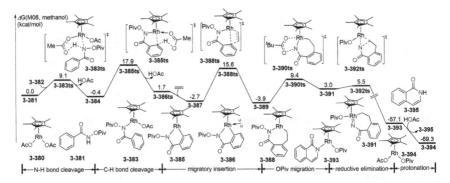

Fig. 3.71 Free energy profiles for Rh(III)-catalyzed (4 + 2) annulation of *N*-acyloxy benzamide with ethylene. The values are the relative free energies given in kcal/mol calculated at the M06/6-311+G(d,p)/SDD//M06/6-31G(d)/LANL2DZ level of theory in methanol

species **3-389**. The followed pivalate migration from N to Rh takes place via a five-membered ring transition state **3-390ts** with an energy barrier of 13.3 kcal/mol, which provide a Rh–nitrene complex **3-391**. The subsequent nitrene insertion into C(aryl)-Rh bond takes place rapidly via three-membered ring-type transition state **3-392ts** with an energy barrier of only 2.5 kcal/mol to result an amido-Rh(III) intermediate **3-393**. Then annulation product can be yielded by protonolysis of acetic acid. The calculated energy barrier for the key step of pivalate migration (13.3 kcal/mol) is closed to that for reductive elimination process in other cases indicating that it would be an alternative pathway.

In 2015, Houk and co-workers [174] performed density functional theory (DFT) calculation to investigate the origin of the reactivity and site-selectivity differences between nicotinamide and its *N*-oxide derivative substrates in Rh(III)-catalyzed annulations with alkynes. As shown in Scheme 3.72, nicotinamide undergoes annulation reaction exhibiting only modest site selectivity preferring second position of the pyridine [175]. In contrast, using its *N*-oxide derivative gives the desired annulation product with high site selectivity at second position of the *N*-oxide pyridine ring.

DFT calculations revealed that in [4+2] annulations of both nicotinamide and its *N*-oxide derivative with substituted acetylene, reductive elimination–oxidative addition pathway is favorable when seven-membered rhodacycle is formed through sequential N–H cleavage, C–H activation, and acetylene insertion. When nicotinamide is used as substrate, the rate-determining step for the C2 annulation is considered as C(aryl)–H bond activation via transition state **3-396ts** with a relative free energy of 25.4 kcal/mol. The insertion of acetylene can occur via transition state **3-397ts**, whose relative free energy is 1.1 kcal/mol lower than that of **3-396ts**. In C4 annulation case, the relative free energy of acetylene insertion transition state **3-399ts** is 25.5 kcal/mol, which is 1.2 kcal/mol higher than that of the corresponding insertion transition state **3-397ts**. The energy difference could be attributed to the extra steric repulsion between C5–H and substituent of acetylene. The relative free energy

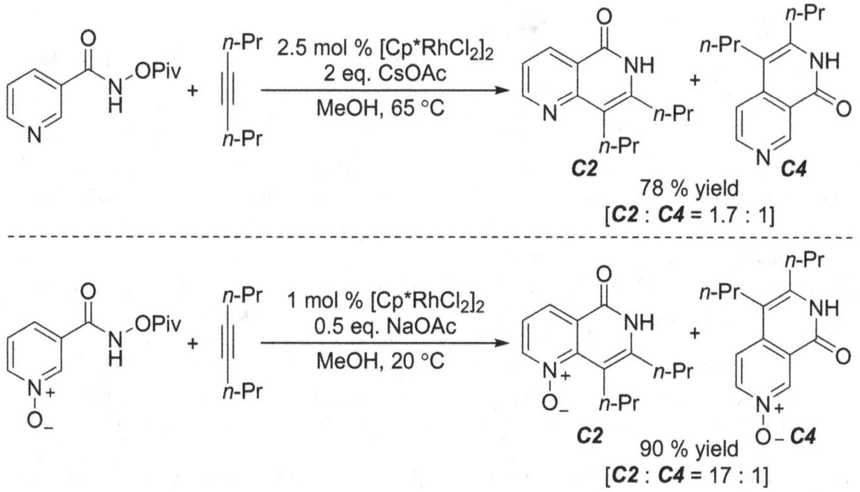

Scheme 3.72 Rh(III)-catalyzed redox-neutral C–H bond activation and annulation of pyridine and pyridine *N*-oxide with alkynes

of **3-396ts** and **3-399ts** is closed, which indicates a poor site-selectivity for annulation. When nicotinamide *N*-oxide is used as substrate, the observable activation energy for both C2 and C4 annulations is significantly reduced because the electron density of pyridine *N*-oxide is lower than pyridine, which leads to the apparently stronger Rh–amide interaction. Therefore, the rate-determining step for either C2 or C4 annulations is acetylene insertion. The relative free energy of transition state **3-401ts** leading to C2 annulation is 2.3 kcal/mol lower than that of **3-403ts**, which causes a C2 site selectivity (Fig. 3.73).

Compared with the widely use of oxidizing directing groups involving N–O bond in Rh(III)-catalyzed C–H activation reactions, the oxidizing directing groups bearing a N–N bond have been relatively less investigated. Zhang and co-workers [160] reported a Rh(III)-catalyzed redox-neutral C–H bond activation and annulation reaction of pyrazolonyl arenes with alkynes, which can be used to direct synthesize *N*-substituted indoles. In this transformation, the cleavage of the covenant N–N bond in pyrazolonyl directing group was performed to turn over the Rh(III) catalyst (Scheme 3.74).

The theoretical calculations were taken by Bi and co-workers [176] to investigate the detailed mechanism for the above annulation. The free energy profiles for the reductive elimination–oxidative addition type pathway are given in Fig. 3.75, which starts from an acetate Rh(III) complex **3-404**. The coordination of pyrazolonyl group onto Rh(III) center increase the acidity of C4-H, which can be deprotonated by acetate via transition state **3-406ts** with an energy barrier of 19.7 kcal/mol to afford a directing covalent N–Rh bond. Then C–H bond activation proceeds via CMD-type transition state **3-408ts** to generate the five-membered rhodacycle **3-409**. The energy barrier for the C–H bond activation is 17.4 kcal/mol. The intermolecular insertion of alkyne into

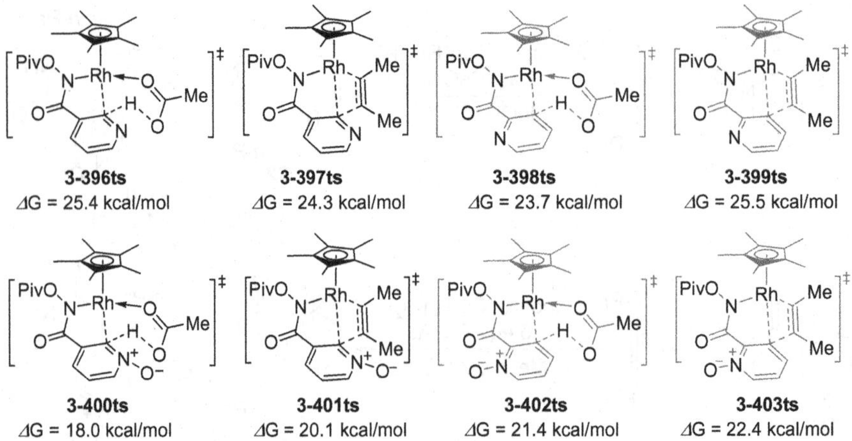
3-396ts
ΔG = 25.4 kcal/mol
3-397ts
ΔG = 24.3 kcal/mol
3-398ts
ΔG = 23.7 kcal/mol
3-399ts
ΔG = 25.5 kcal/mol
3-400ts
ΔG = 18.0 kcal/mol
3-401ts
ΔG = 20.1 kcal/mol
3-402ts
ΔG = 21.4 kcal/mol
3-403ts
ΔG = 22.4 kcal/mol

Fig. 3.73 The site-selectivity for Rh(III)-catalyzed redox-neutral C–H bond activation and annulation of nicotinamide and its *N*-oxide derivative with alkynes. The values are the relative free energies given in kcal/mol calculated at the M06/6-311+G(d,p)/SDD//M06/6-31G(d)/LANL2DZ level of theory in methanol

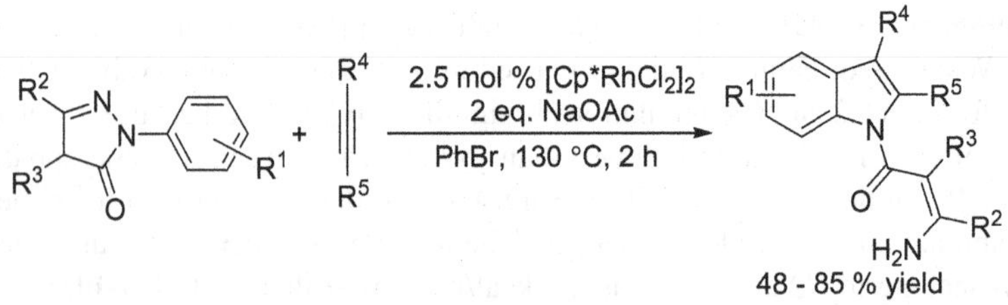
2.5 mol % [Cp*RhCl2]2
2 eq. NaOAc
PhBr, 130 °C, 2 h
48 - 85 % yield

Scheme 3.74 Rh(III)-catalyzed redox-neutral C–H bond activation and annulation reaction of pyrazolones with alkynes

the Rh–C bond rather than the Rh–N bond occurs via transition state **3-410ts** with an energy barrier of 10.4 kcal/mol leading to the extension of rhodacycle in **3-411**. The following reductive elimination takes place via a four-membered ring-type vinyl migration transition state **3-412ts** with a barrier of 21.5 kcal/mol to give a zwitterionic Rh(I) intermediate **3-413**. Then an N–N bond cleavage occurs via oxidative addition transition state **3-414ts** involves a free energy barrier of 15.7 kcal/mol to give the Rh(III)-nitrene complex **3-415**. After the N–N bond cleavage, the desired indole product **3-416** is obtained and the active catalyst **3-404** is regenerated through two steps of protonolysis of the N atom by two acetic acids.

In another example, Lin and co-workers [177] reported a theoretical study for the Rh(III)-catalyzed C–H activation and annulation reaction of acetylhydrazinyl arenes with alkynes to prepare indoles, which is developed by Glorius and co-workers [178] in 2013. In this reaction, acetylhydrazinyl group is considered as the internal oxidizing directing group (Scheme 3.76).

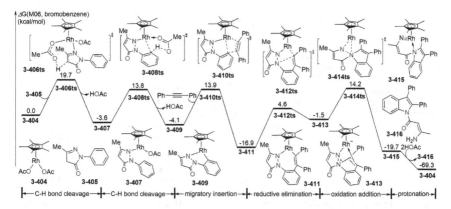

Fig. 3.75 Free energy profiles for Rh(III)-catalyzed redox-neutral C–H bond activation and annulation reaction of pyrazolones with alkynes. The values are the relative free energies given in kcal/mol calculated at the M06/6-31G(d,p)/LANL2DZ//B3-LYP/6-31G(d,p)/LANL2DZ level of theory in methanol

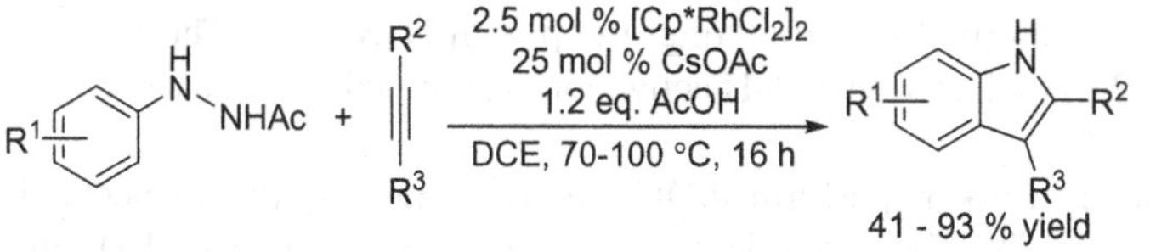

Scheme 3.76 Rh(III)-catalyzed C–H activation and annulation reaction of 2-acetyl-1-arylhydrazines with alkynes

As shown in Fig. 3.77, a reductive elimination–oxidative addition pathway was considered theoretically, which starts from Cp*Rh(OAc)$_2$ **3-417**. The deprotonation of the metal-activated N–H bond in acetylhydrazinyl leads to the formation of the N–Rh bond in intermediate **3-420**. The following CMD type C–H bond activation proceeds through the six-membered-ring transition state **3-421ts** with an overall energy barrier of 25.9 kcal/mol to give the five-membered rhodacycle **3-422**. The coordinated acetylene would insert into the Rh–C(aryl) bond via transition state **3-423ts** to form the seven-membered rhodacycle **3-424**. The subsequent Rh–H exchange generates six-membered rhodacycle complex **3-425**, which is 5.5 kcal/mol more stable than **3-424**. Then, the C–N reductive elimination occurs via transition state **3-426ts** to form Rh(I) intermediate **3-427** requires overcoming an energy barrier of 19.5 kcal/mol. Then **3-427** undergoes oxidation addition to break the N–N bond via transition state **3-428ts** with an energy barrier of 19.1 kcal/mol to form indolyl Rh(III) **3-429**, which can be protonated to yield the desired indole product **3-430** and regenerate the Rh(III) species **3-417** to complete the catalytic cycle.

The oxidizing N–S directing groups are also valuable as internal oxidant, though the cleavable N–S bonds have been rarely reported as oxidant. In 2016, Li and Lan [179] developed a Rh(III)-catalyzed C–H bond activation and annulation reaction

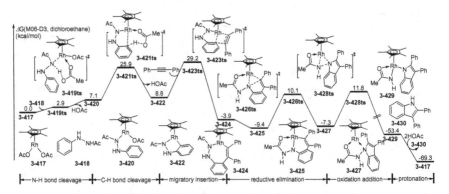

Fig. 3.77 Free energy profiles for Rh(III)-catalyzed C–H activation and annulation reaction of 2-acetyl-1-arylhydrazines with alkynes. The values are the relative free energies given in kcal/mol calculated at the M06/6-31G(d)/LANL2DZ//M06/6-31G(d)/LANL2DZ level of theory in dichloroethane

of N-sulfinyl imines with olefins resulting isoindole derivatives. This transformation is assisted by an oxidizing directing group with N–S bond. The KIE experiments ($k_H/k_D = 2.3$) indicated that C–H activation is involved in the turnover-limiting step (Scheme 3.78).

Lan and co-workers performed DFT studies to investigate the mechanism of this reaction (Fig. 3.79). The catalytic cycle starts with the cationic Rh(III) active catalyst **3-431**. The acetate assisted C–H bond activation occurs via transition state **3-433ts** to generate aryl-Rh(III) intermediate **3-434** with an energy barrier of 27.7 kcal/mol. The subsequent migratory insertion of ethyl acrylate takes place via transition state **3-436ts** with an activation free energy of 20.0 kcal/mol to reversibly produce the seven-membered rhodacycle **3-437**. The β-hydride elimination of **3-437** is established via transition state **3-438ts** with a barrier of 19.9 kcal/mol. Then, the N–S bond cleavage occurs upon intramolecular abstraction of the Rh–H proton by the sulfinyl oxygen via transition state **3-440ts** to afford the imido-Rh(III) **3-441**, which can be considered as a metathesis type hydride oxidation. The migratory insertion of coordinated olefin into N-Rh(III) bond takes place via transition state **3-442ts** with a barrier of only 13.5 kcal/mol. The following protonation of the enolate carbon moiety and deprotonation of the methine C–H bond forms isoindolyl Rh(III) **3-444**.

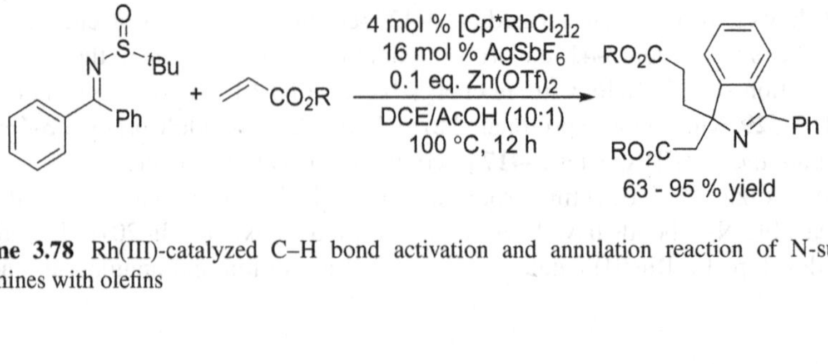

Scheme 3.78 Rh(III)-catalyzed C–H bond activation and annulation reaction of N-sulfinyl ketoimines with olefins

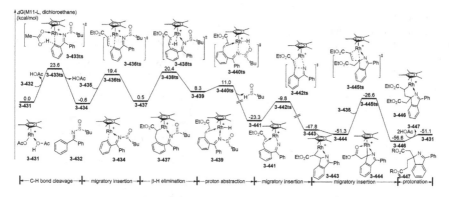

Fig. 3.79 Free energy profiles for Rh(III)-catalyzed C–H bond activation and annulation reaction of N-sulfinyl ketoimines with olefins. The values are the relative free energies given in kcal/mol calculated at the M11-L/6-311+G(d)/LANL08(f)//B3-LYP/6-31G(d)/LANL08(f) level of theory in dichloroethane

The migratory insertion of the second ethyl acrylate proceeds via transition state **3-445ts** to produce the N-bound isoindole complex **3-446**. The overall activation free energy for **3-445ts** is 24.7 kcal/mol. The protonation and ligand exchange release the product **3-447** and regenerate the active Rh(III) catalyst **3-431** to complete the catalytic cycle.

3.7 Rh-Mediated Carbene Transfer

Carbenes can be considered as an unsaturated 6e carbon, which is a popular active intermediate in organic synthesis for the construction of new C–C bonds [180–183]. The unsaturation character of carbenes usually reveals insertion reactivity, therefore, carbene can connect with a nucleophile and an electrophile directly onto one carbon atom. In organometallic catalysis, carbene can be stabilized by the coordination onto transition metal in the formation of metal-carbene complexes [184–187]. The coordinated carbene used a sp^2 hybrid orbital bond with metal, while, the back donation d-p orbital exhibits electrophilicity. Therefore, a metal-carbene can react with a C–H bond by the outer-sphere carbene insertion into this bond to achieve C–H activation. Alternatively, when C–H activation takes place first, the carbene moiety also can insert into metal–carbon bond through an inner-sphere process to achieve C–H functionalization. As shown in Scheme 3.80, when carbenation takes place first to afford Rh–carbene complex **3-449**, the electrophilic carbene can insert into C–H bond through a concerted outer-sphere process to achieve HOMO activation of C–H bond. In an alternative catalytic cycle, a CMD type C–H activation provides an aryl-Rh species, which can be carbenated to afford aryl Rh–carbene complex **4-454**. The followed migratory insertion of carbene into C(aryl)-Rh bond results the C(aryl)-C(carbene) coupling. The final product can be yielded by a protonolysis.

Scheme 3.80 General
mechanism for Rh-mediated
insertion of carbene into
C–H bonds

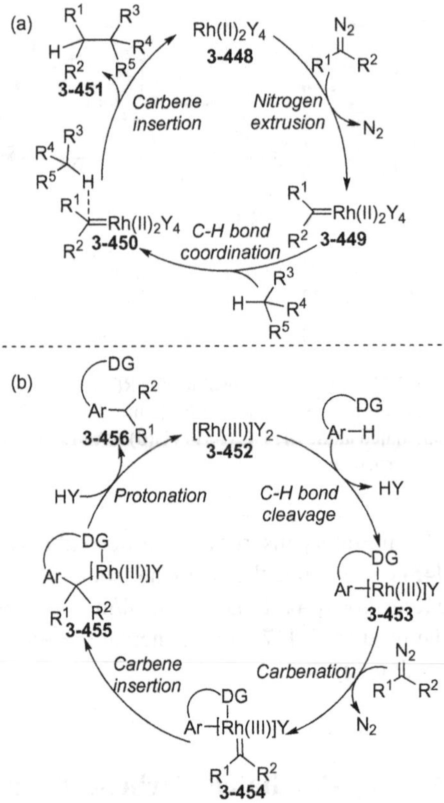

3.7.1   Outer-Sphere C–H Activation by Rh–Carbene Complex

When Rh–carbene complex is formed first, the electrophilicity of carbene leads to a
HOMO activation of C–H bond, which can undergo an outer-sphere carbene insertion
into C–H bond to achieve C–H functionalization through either concerted or stepwise
pathways [188–191].

Yamanaka and co-workers [192] performed theoretical study to understand the
detailed mechanism for the $Rh_2(O_2CH)_4$-catalyzed C–H bond activation undergoing
carbene insertion. Diazo compound **3-458** was used as carbeniod, which can coor-
dinate onto dirhodium species **3-457** to form complex **3-459** by 19.9 kcal/mol exergonic. Carbenation then occurs via transition state **3-460ts** with an energy barrier
of 16.6 kcal/mol to form Rh–carbene complex **3-461**. The decomposition of diazo
compound is 16.8 kcal/mol endergonic in the presence of dirhodium. When methane
is used as substrate, the intermolecular carbene insertion into C(alkyl)–H bond takes
place via a three-membered ring transition state **3-464ts** with an energy barrier of
5.9 kcal/mol to afford ethane directly. The reactivity of propane was also considered theoretically. An almost barrierless process via three-membered ring transition

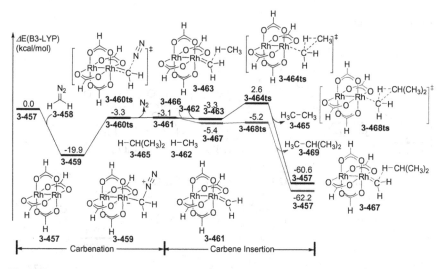

Fig. 3.81 Free energy profiles for $Rh_2(OAc)_4$-catalyzed C–H bond activation and C–C bond formation reaction of diazomethane with methane and propane. The values are the relative energies given in kcal/mol calculated at the B3-LYP/6-31G(d)/LANL2DZ level of theory

state **3-468ts** was found to achieve secondary C(alkyl)–H bond activation. Computational results clearly revealed that the carbene insertion into secondary C–H bond is preferred (Fig. 3.81).

The outer-sphere C–H activation through carbene insertion can occur through a stepwise process depending on the nucleophilicity of C–H bond. In 2015, Xie and co-workers [193] reported a DFT study on the mechanism of dirhodium-catalyzed carbene insertion into the nucleophilic C–H bond of indole. The calculated rational free energy profiles for the stepwise carbene insertion are shown in Fig. 3.82. Carbenation of dirhodium by diazo compound can afford Rh–carbene complex **3-470**, which can undergo a nucleophilic attack by C3 of indole via transition state **3-472ts** to afford a zwitterionic intermediate **3-473**. The subsequent intramolecular [1, 4] proton transfer from C to O proceeds via transition state **3-474ts** to generate an enol coordinated dirhodium complex **3-475**. Releasing of insertion product **3-477** regenerates dirhodium compound **3-476** with 17.8 kcal/mol exothermic.

3.7.2 Sequential C–H Activation and Inner-Sphere Carbene Insertion

In a CMD type C(aryl)-H activation by using Rh(III) species, the provided C(aryl)-Rh bond can react with in situ generated carbene by an inner-sphere insertion for the construction of new C–C bond. In 2015, Li et al. [194] reported an experimental and theoretical study of Rh(III)-catalyzed C–H bond activation of phenacyl ammonium

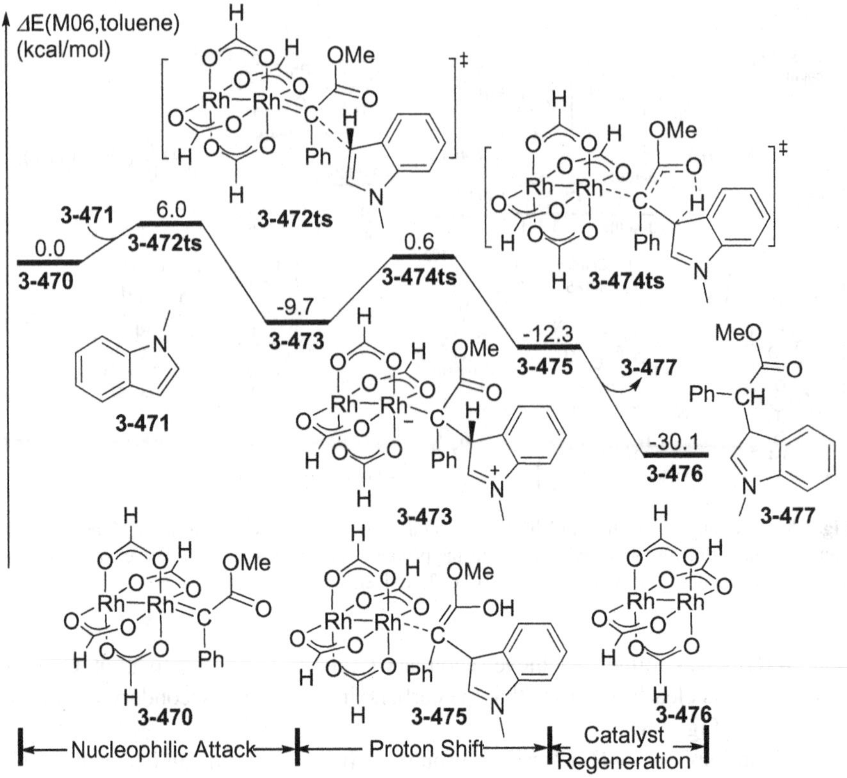

Fig. 3.82 Free energy profiles for Rh(II)-catalyzed carbene insertion into the C–H bond of indole. The values are the relative energies given in kcal/mol calculated at the M06/6-31+G(d,p)/LANL2DZ//B3-LYP/6-31G(d)/LANL2DZ level of theory in toluene

salts in the presence of diazo compounds. In this reaction, ammonium moiety was used as leaving group to keep redox neutral. The KIE experiments ($k_H/k_D = 4.6$) indicated that the C–H cleavage is involved in the rate-limiting step (Scheme 3.83).

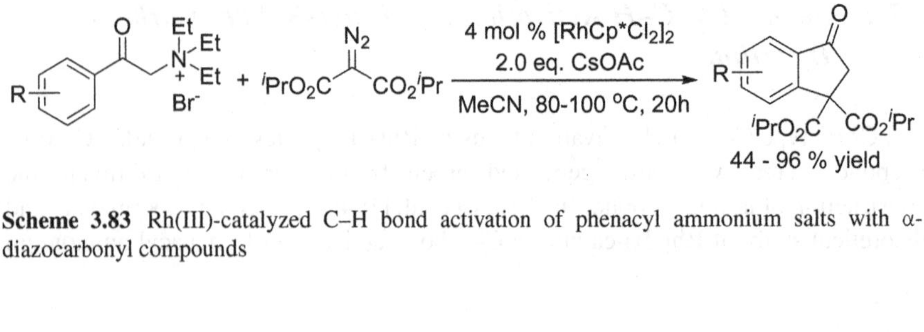

Scheme 3.83 Rh(III)-catalyzed C–H bond activation of phenacyl ammonium salts with α-diazocarbonyl compounds

As shown in Fig. 3.84, phenacyl ammonium **3-479** can react with active species acetate Rh(III) **3-478** to afford a C-bound enolate Rh(III) **3-480** through deprotonation achieving covalent chelation. Then CMD type C–H bond activation occurs via transition state **3-481ts** with an energy barrier of 27.2 kcal/mol. The C–H bond activation was considered to be the rate-determining step for the catalytic cycle, which is consistent with the experimental results. The diazo compound **3-483** then coordinates onto generated five-membered rhodacyclic **3-482**, leading to a carbenation through denitrogenation transition state **3-484ts**. The generated carbene can irreversibly insert into C(aryl)-Rh bond via transition state **3-486ts** with a barrier of 10.2 kcal/mol to form six-membered rhodacycle **3-487**. Subsequently, deamination takes place via transition state **3-488ts** to afford a cyclic Rh–carbene complex **3-489**. The sequential intramolecular carbene insertion and protonolysis yield indenone product **3-492** with the regeneration of Rh(III) active species **3-478**.

In another example of Rh(III)-catalyzed inner-sphere C–H activation and carbene insertion reaction, oxidizing directing group oxime ester was chosen to keep redox neutral. Xia and co-workers [195] reported a theoretical study on the mechanism of Rh(III)-catalyzed redox-neutral C–H activation and annulation of *N*-pivaloyloxy benzamides with diazo compounds, which is independently developed by Rovis [196] and Cui [197]. Interestingly, when α,β-unsaturated diazo compounds were used, azepinones were observed as major product (Scheme 3.85).

The calculated detailed free energy profiles for the inner-sphere C–H carbenation are shown in Fig. 3.86, which starts from a covalent directing group loading through deprotonation of benzamide via transition state **3-495ts**. The subsequent C–H activation takes place via CMD-type transition state **3-497ts** with an energy barrier of 17.9 kcal/mol. Then carbenation with α,β-unsaturated diazo compound **3-499** occurs via transition state **3-500ts** irreversibly results a Rh–carbene complex **3-501**. A rapid carbene insertion into C(aryl)-Rh bond via transition state **3-502ts** results a six-membered rhodacycle **3-503**. The isomerization of allylic Rh moiety extends

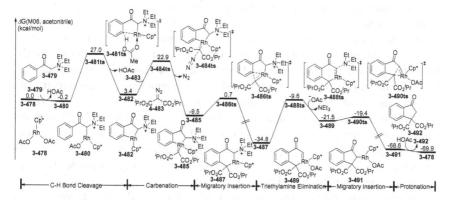

Fig. 3.84 Free energy profiles for Rh(III)-catalyzed C–H bond activation of phenacyl ammonium salts with α-diazocarbonyl. The values are the relative energies given in kcal/mol calculated at the M06/6-311+G(d)/SDD//B3-LYP/6-31+G(d)/SDD level of theory in acetonitrile

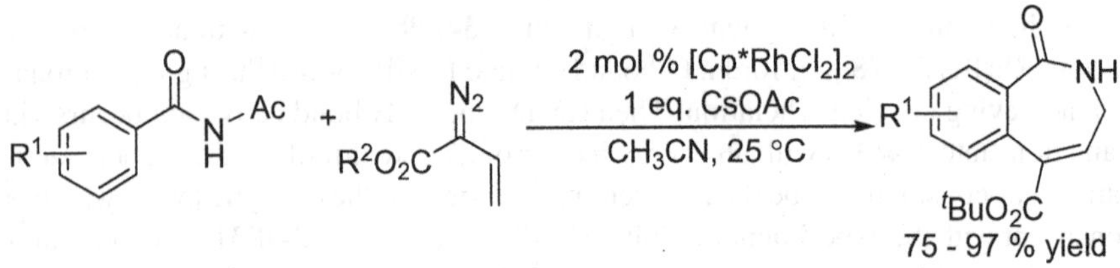

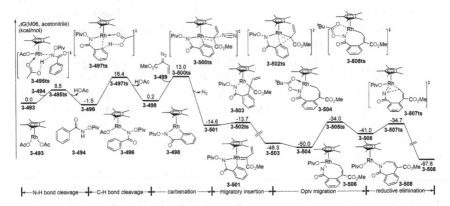

Scheme 3.85 Rh(III)-catalyzed redox-neutral C–H activation and annulation of benzamides and diazo compounds

Fig. 3.86 Free energy profiles for Rh(III)-catalyzed redox-neutral C–H activation and annulation of benzamides and diazo compounds. The values are the relative energies given in kcal/mol calculated at the M06/6-311+G(d,p)/SDD//M06/6-31G(d)/LANL2DZ level of theory in acetonitrile

rhodacycle to eight-membered ring in intermediate **3-504**. A pivalate migration from N to Rh affords a Rh–nitrene complex, which can undergo a nitrene insertion to result azepinone product by the followed protonolysis.

In a carbene involved annulation, an endogenous oxidant is necessary to keep redox neutral. Thereinto, amine N-oxide can be a good choice, which also plays as a directing group in Rh-catalyzed C–H activation and annulation with carbenoids. Zhou and co-workers [198] reported a Rh(III)-catalyzed redox-neutral annulation of N-alkyl aminonaphthalene N–oxides with diazo compounds. This reaction represents the first example of the dual functionalization of unactivated primary C(sp^3)–H and C(sp^2)–H bonds with diazo compounds. The KIE experiment of C(sp^3)–H bond ($k_H/k_D = 1$) indicated that the C(sp^3)–H bond cleavage is not involved in the rate-determining step in the catalytic cycle. In addition, the KIE experiment of C(sp^2)–H bond ($k_H/k_D = 3$) shown that the *ortho* C(sp^2)–H bond cleavage might be related to the rate-limiting step (Scheme 3.87).

Simultaneously, Zhou and co-workers [198] performed DFT calculations to gain further insight into the detailed mechanism for the above reaction. As shown in Fig. 3.88, an acetate-assisted C(sp^2)–H bond activation directing by N-oxide group

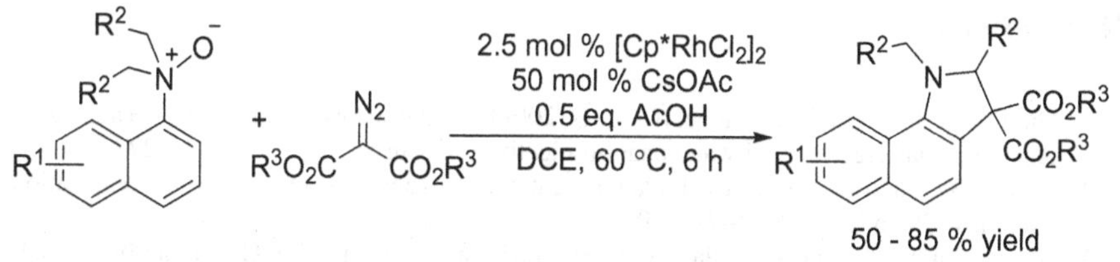

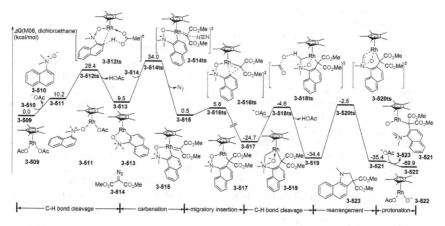

Scheme 3.87 Rh(III)-catalyzed redox-neutral annulation of 1-naphthylamine *N*-oxides with diazo compounds

Fig. 3.88 Free energy profiles for Rh(III)-catalyzed redox-neutral annulation of 1-naphthylamine *N*-oxides with diazo compounds. The values are the relative energies given in kcal/mol calculated at the M06/6-31+G(d)/SDD//M06/6-311+G(d)/SDD level of theory in dichloroethane

occurs via transition state **3-512ts** with an overall activation energy of 28.4 kcal/mol. The coordination of diazo compound leads to denitrogenation via transition state **3-514ts** resulting in a cyclic Rh–carbene complex **3-515** by the release of gaseous dinitrogen. Then carbene insertion into C(aryl)–Rh bond forms a six-membered rhodacycle **3-517**. An outer-sphere acetate-assisted deprotonation of α-C(alkyl)-H bond takes place via transition state **3-518ts**. In the presence of neighboring formal cationic nitrogen, the energy barrier for the deprotonation process is only 20.1 kcal/mol. The following intramolecular oxo transfer takes place via transition state **3-520ts** with an energy barrier of 31.9 kcal/mol to form a zwitterionic Rh(III)-oxo complex **3-521**. An outer-sphere C–C coupling results indole product **3-523** with the generation of Rh(III)-oxo complex **3-522**, which can react with acetic acid to regenerate active species **3-509**.

References

1. Giri R, Shi BF, Engle KM, Maugel N, Yu JQ (2009) Transition metal-catalyzed C–H activation reactions: diastereoselectivity and enantioselectivity. Chem Soc Rev 38:3242–3272
2. Lyons TW, Sanford MS (2010) Palladium-catalyzed ligand-directed C–H functionalization reactions. Chem Rev 110:1147–1169
3. Ackermann L (2011) Carboxylate-assisted transition-metal-catalyzed C–H bond functionalizations: mechanism and scope. Chem Rev 111:1315–1345
4. Bergman RG (2007) Organometallic chemistry: C–H activation. Nature 446:391–393
5. Godula K, Sames D (2006) C–H bond functionalization in complex organic synthesis. Science 312:67–72
6. Gunay A, Theopold KH (2010) C–H bond activations by metal oxo compounds. Chem Rev 110:1060–1081
7. Kakiuchi F, Kochi T (2008) Transition-Metal-Catalyzed Carbon–Carbon bond formation via Carbon–Hydrogen bond cleavage. Synthesis 2008:3013–3039
8. Dateer RB, Chang S (2015) Selective cyclization of arylnitrones to indolines under external oxidant-free conditions: dual role of Rh(III) catalyst in the C–H activation and oxygen atom transfer. J Am Chem Soc 137:4908–4911
9. Lewis JC, Bergman RG, Ellman JA (2008) Direct functionalization of nitrogen heterocycles via Rh-catalyzed C–H bond activation. Acc Chem Res 41:1013–1025
10. Lin Y, Zhu L, Lan Y, Rao Y (2015) Development of a Rhodium(II)-Catalyzed chemoselective C(sp(3))–H Oxygenation. Chem Eur J 21:14937–14942
11. Qin X, Li X, Huang Q, Liu H, Wu D, Guo Q, Lan J, Wang R, You J (2015) Rhodium(III)-catalyzed ortho C–H heteroarylation of (hetero)aromatic carboxylic acids: a rapid and concise access to pi-conjugated poly-heterocycles. Angew Chem Int Ed 54:7167–7170
12. Shin K, Kim H, Chang S (2015) Transition-metal-catalyzed C–N bond forming reactions using organic azides as the nitrogen source: a journey for the mild and versatile C–H amination. Acc Chem Res 48:1040–1052
13. Song G, Li X (2015) Substrate activation strategies in rhodium(III)-catalyzed selective functionalization of arenes. Acc Chem Res 48:1007–1020
14. Tan KL, Bergman RG, Ellman JA (2002) Intermolecular coupling of isomerizable alkenes to heterocycles via rhodium-catalyzed C–H bond activation. J Am Chem Soc 124:13964–13965
15. Yu S, Li Y, Kong L, Zhou X, Tang G, Lan Y, Li X (2016) Mild acylation of $C(sp3)$–H and $C(sp^2)$–H bonds under redox-neutral Rh(III) catalysis. ACS Catal 6:7744–7748
16. Yu S, Tang G, Li Y, Zhou X, Lan Y, Li X (2016) Anthranil: an aminating reagent leading to bifunctionality for both C(sp(3))-H and C(sp(2))-H under Rhodium(III) catalysis. Angew Chem Int Ed 55:8696–8700
17. Colby DA, Bergman RG, Ellman JA (2010) Rhodium-catalyzed C–C bond formation via heteroatom-directed C–H bond activation. Chem Rev 110:624–655
18. Gridnev ID, Imamoto T (2004) On the mechanism of stereoselection in Rh-catalyzed asymmetric hydrogenation: a general approach for predicting the sense of enantioselectivity. Acc Chem Res 37:633–644
19. Davies DL, Macgregor SA, McMullin CL (2017) Computational studies of carboxylate-assisted C–H activation and functionalization at group 8–10 transition metal centers. Chem Rev 117:8649–8709
20. Sperger T, Sanhueza IA, Kalvet I, Schoenebeck F (2015) Computational studies of synthetically relevant homogeneous organometallic catalysis involving Ni, Pd, Ir, and Rh: an overview of commonly employed DFT methods and mechanistic insights. Chem Rev 115:9532–9586
21. Ackermann L, Rn Vicente, Kapdi AR (2009) Transition-metal-catalyzed direct arylation of (hetero)arenes by C–H bond cleavage. Angew Chem Int Ed 48:9792–9826
22. Bedford RB, Coles SJ, Hursthouse MB, Limmert ME (2003) The catalytic intermolecular orthoarylation of phenols**. Angew Chem Int Ed 42:112–114

23. Lewis JC, Berman AM, Bergman RG, Ellman JA (2008) Rh(I)-catalyzed arylation of heterocycles via C–H bond activation: expanded scope through mechanistic insight. J Am Chem Soc 130:2493–2500
24. Wang X, Lane BS, Sames D (2005) Direct C-arylation of free (NH)-indoles and pyrroles catalyzed by Ar-Rh(III) complexes assembled in situ. J Am Chem Soc 127:4996–4997
25. Santoro S, Himo F (2018) Mechanism and selectivity of rhodium-catalyzed C–H bond arylation of indoles. Int J Quantum Chem 118:e25526
26. Oi S, Fukita S, Inoue Y (1998) Rhodium-catalysed direct ortho arylation of 2-arylpyridines with arylstannanes via C–H activation. Chem Comm:2439–2440
27. Ueura K, Satoh T, Miura M (2005) Rhodium-catalyzed arylation using arylboron compounds: efficient coupling with aryl halides and unexpected multiple arylation of benzonitrile. Org Lett 7:2229–2231
28. Miyamura S, Tsurugi H, Satoh T, Miura M (2008) Rhodium-catalyzed regioselective arylation of phenylazoles and related compounds with arylboron reagents via C–H bond cleavage. J Organomet Chem 693:2438–2442
29. Vogler T, Studer A (2008) Oxidative coupling of arylboronic acids with arenes via Rh-catalyzed direct C–H arylation. Org Lett 10:129–131
30. Karthikeyan J, Haridharan R, Cheng CH (2012) Rhodium(III)-catalyzed oxidative C–H coupling of N-methoxybenzamides with aryl boronic acids: one-pot synthesis of phenanthridinones. Angew Chem Int Ed 51:12343–12347
31. Zheng J, Zhang Y, Cui S (2014) Rh(III)-catalyzed selective coupling of N-methoxy-1H-indole-1-carboxamides and aryl boronic acids. Org Lett 16:3560–3563
32. Wencel-Delord J, Nimphius C, Patureau FW, Glorius F (2012) [Rh(III)Cp*]-catalyzed dehydrogenative aryl-aryl bond formation. Angew Chem Int Ed 51:2247–2251
33. Zhao D, Li X, Han K, Li X, Wang Y (2015) Theoretical investigations on Rh(III)-catalyzed cross-dehydrogenative aryl-aryl coupling via C–H bond activation. J Phys Chem A 119:2989–2997
34. Colby DA, Tsai AS, Bergman RG, Ellman JA (2012) Rhodium catalyzed chelation-assisted C–H bond functionalization reactions. Acc Chem Res 45:814–825
35. He Q, Yamaguchi T, Chatani N (2017) Rh(I)-Catalyzed Alkylation of ortho-C–H Bonds in Aromatic Amides with Maleimides. Org Lett 19:4544–4547
36. Song G, Wang F, Li X (2012) C–C, C–O and C–N bond formation via rhodium(III)-catalyzed oxidative C–H activation. Chem Soc Rev 41:3651–3678
37. Tran G, Hesp KD, Mascitti V, Ellman JA (2017) Base-controlled completely selective linear or branched Rhodium(I)-Catalyzed C–H ortho-alkylation of azines without preactivation. Angew Chem Int Ed 56:5899–5903
38. Deng H, Li H, Wang L (2015) A unique alkylation of azobenzenes with allyl acetates by Rh(III)-catalyzed C–H functionalization. Org Lett 17:2450–2453
39. Huang L, Wang Q, Qi J, Wu X, Huang K, Jiang H (2013) Rh(III)-catalyzed ortho-oxidative alkylation of unactivated arenes with allylic alcohols. Chem Sci 4:2665
40. Lewis JC, Bergman RG, Ellman JA (2007) Rh(I)-catalyzed alkylation of quinolines and pyridines via C–H bond activation. J Am Chem Soc 129:5332–5333
41. Shibata K, Chatani N (2014) Rhodium-catalyzed alkylation of C-H bonds in aromatic amides with alpha, beta-unsaturated esters. Org Lett 16:5148–5151
42. Sharma U, Park Y, Chang S (2014) Rh(III)-catalyzed traceless coupling of quinoline N-oxides with internal diarylalkynes. J Org Chem 79:9899–9906
43. Zhang X, Qi Z, Li X (2014) Rhodium(III)-catalyzed C–C and C–O coupling of quinoline N-oxides with alkynes: combination of C–H activation with O-atom transfer. Angew Chem Int Ed 53:10794–10798
44. Chen X, Zheng G, Li Y, Song G, Li X (2017) Rhodium-catalyzed site-selective coupling of indoles with diazo esters: C4-alkylation versus C2-annulation. Org Lett 19:6184–6187
45. Wu Y, Chen Z, Yang Y, Zhu W, Zhou B (2018) Rh(III)-catalyzed redox-neutral unsymmetrical C–H alkylation and amidation reactions of N-phenoxyacetamides. J Am Chem Soc 140:42–45

46. Wiedemann SH, Ellman JA, Bergman RG (2006) Rhodium-catalyzed direct C–H addition of 3,4-dihydroquinazolines to alkenes and their use in the total synthesis of vasicoline. J Org Chem 71:1969–1976

47. Wiedemann SH, Lewis JC, Ellman JA, Bergman RG (2006) Experimental and computational studies on the mechanism of N-heterocycle C–H activation by Rh(I). J Am Chem Soc 128:2452–2462

48. Tan KL, Bergman RG, Ellman JA (2001) Annulation of alkenyl-substituted heterocycles via rhodium-catalyzed intramolecular C–H activated coupling reactions. J Am Chem Soc 123:2685–2686

49. Kwak J, Ohk Y, Jung Y, Chang S (2012) Rollover cyclometalation pathway in rhodium catalysis: dramatic NHC effects in the C–H bond functionalization. J Am Chem Soc 134:17778–17788

50. Dong FMaG (2014) Regioselective ketone a-alkylation with simple olefins via dual activation. Science 345:68–72

51. Park YJ, Park JW, Jun CH (2008) Metal-organic cooperative catalysis in C–H and C–C bond activation and its concurrent recovery. Acc Chem Res 41:222–234

52. Ko HM, Dong G (2014) Cooperative activation of cyclobutanones and olefins leads to bridged ring systems by a catalytic [4+2] coupling. Nat Chem 6:739–744

53. Dang Y, Qu S, Tao Y, Deng X, Wang Z-X (2015) Mechanistic insight into ketone alpha-alkylation with unactivated olefins via C–H activation promoted by metal-organic cooperative catalysis (MOCC): enriching the MOCC chemistry. J Am Chem Soc 137:6279–6291

54. Zhang X, Qi Z, Li X (2014) Rhodium(III)-Catalyzed C–C and C–O coupling of quinolinen-Oxides with alkynes: combination of C–H activation with O-atom transfer. Angew Chem Int Ed 126:10970–10974

55. Li Y, Liu S, Qi Z, Qi X, Li X, Lan Y (2015) The mechanism of N–O bond cleavage in rhodium-catalyzed C–H bond functionalization of quinoline N-oxides with alkynes: a computational study. Chem Eur J 21:10131–10137

56. Li Y, Shan C, Yang YF, Shi F, Qi X, Houk KN, Lan Y (2017) Mechanism, regio-, and diastereoselectivity of Rh(III)-catalyzed cyclization reactions of N-arylnitrones with alkynes: a density functional theory study. J Phys Chem A 121:4496–4504

57. Xing YY, Liu JB, Tian YY, Sun CZ, Huang F, Chen DZ (2016) Computational mechanistic study of redox-neutral Rh(III)–catalyzed C–H activation reactions of arylnitrones with alkynes: role of noncovalent interactions in controlling selectivity. J Phys Chem A 120:9151–9158

58. Jeong J, Patel P, Hwang H, Chang S (2014) Rhodium(III)-catalyzed C–C bond formation of quinoline N-oxides at the C-8 position under mild conditions. Org Lett 16:4598–4601

59. Deng C, Lam WH, Lin Z (2017) Computational studies on Rhodium(III) catalyzed C–H functionalization versus deoxygenation of quinoline N-oxides with diazo compounds. Organometallics 36:650–656

60. Brasse M, Cámpora J, Ellman JA, Bergman RG (2013) Mechanistic study of the oxidative coupling of styrene with 2-phenylpyridine derivatives catalyzed by cationic Rhodium(III) via C–H activation. J Am Chem Soc 135:6427–6430

61. Chen J, Song G, Pan CL, Li X (2010) Rh(III)-catalyzed oxidative coupling of N-Aryl-2-aminopyridine with alkynes and alkenes. Org Lett 12:5426–5429

62. Karthikeyan J, Yoshikai N (2014) Rhodium(III)-catalyzed directed peri-C-H alkenylation of anthracene derivatives. Org Lett 16:4224–4227

63. Chang WTT, Smith RC, Regens CS, Bailey AD, Werner NS, Denmark SE, Denmark SE, Liu JHC (2011) Cross-coupling with organosilicon compounds silicon-based cross-coupling reactions in the total synthesis of natural products. Org React 49:213-746-2986

64. Li X, Gong X, Zhao M, Song G, Deng J, Li X (2011) Rh(III)-catalyzed oxidative olefination of N-(1-Naphthyl)sulfonamides using activated and unactivated alkenes. Org Lett 13:5808–5811

65. Lin W, Li W, Lu D, Su F, Wen TB, Zhang HJ (2018) Dual effects of cyclopentadienyl ligands on Rh(III)-catalyzed dehydrogenative arylation of electron-rich alkenes. ACS Catal 8:8070–8076

66. Satoh T, Miura M (2010) Oxidative coupling of aromatic substrates with alkynes and alkenes under rhodium catalysis. Chem Eur J 16:11212–11222
67. Takahama Y, Shibata Y, Tanaka K (2015) Oxidative olefination of anilides with unactivated alkenes catalyzed by an (electron-deficient η5-cyclopentadienyl)Rhodium(iii) complex under ambient conditions. Chem -Eur J 21:9053–9056
68. Tsai AS, Brasse M, Bergman RG, Ellman JA (2011) Rh(III)-catalyzed oxidative coupling of unactivated alkenes via C–H activation. Org Lett 13:540–542
69. Xue X, Xu J, Zhang L, Xu C, Pan Y, Xu L, Li H, Zhang W (2016) Rhodium(III)-catalyzed direct C–H olefination of arenes with aliphatic olefins. Adv Synth Catal 358:573–583
70. Zhao P, Wang F, Han K, Li X (2012) Rhodium(III)-catalyzed cyclization-olefination of n-acetoxyl ketoimine-alkynes. Org Lett 14:3400–3403
71. Zheng J, Cui W-J, Zheng C, You S-L (2016) Synthesis and application of chiral spiro cp ligands in rhodium-catalyzed asymmetric oxidative coupling of biaryl compounds with alkenes. J Am Chem Soc 138:5242–5245
72. Zheng J, You S-L (2014) Construction of axial chirality by rhodium-catalyzed asymmetric dehydrogenative heck coupling of biaryl compounds with alkenes. Angew Chem Int Ed 53:13244–13247
73. Zheng L, Wang J (2012) Direct oxidative coupling of arenes with olefins by Rh-Catalyzed -H activation in air: observation of a strong cooperation of the acid. Chem Eur J 18:9699–9704
74. Zhu W, Gunnoe TB (2020) Advances in rhodium-catalyzed oxidative arene alkenylation. Acc Chem Res 53:920–936
75. Zhu W, Gunnoe TB (2020) Rhodium-catalyzed arene alkenylation using only dioxygen as the oxidant. ACS Catal 10:11519–11531
76. Gong TJ, Xiao B, Liu ZJ, Wan J, Xu J, Luo DF, Fu Y, Liu L (2011) Rhodium-catalyzed selective C–H activation/olefination of phenol carbamates. Org Lett 13:3235–3237
77. Zhang Q, Yu H-Z, Li Y-T, Liu L, Huang Y, Fu Y (2013) Computational study on mechanism of Rh(III)-catalyzed oxidative heck coupling of phenol carbamates with alkenes. Dalton Trans 42:4175–4184
78. Xu L, Zhang C, He Y, Tan L, Ma D (2015) Rhodium-catalyzed regioselective C7-functionalization of N-pivaloylindoles. Angew Chem Int Ed 55:321–325
79. Han L, Ma X, Liu Y, Yu Z, Liu T (2018) Mechanistic insight into the C7-selective C–H functionalization of N-acyl indole catalyzed by a rhodium complex: a theoretical study. Org Chem Front 5:725–733
80. Azpíroz R, Di Giuseppe A, Urriolabeitia A, Passarelli V, Polo V, Pérez-Torrente JJ, Oro LA, Castarlenas R (2019) Hydride-rhodium(iii)-n-heterocyclic carbene catalyst for tandem alkylation/alkenylation via C–H activation. ACS Catal 9:9372–9386
81. Boyarskiy VP, Ryabukhin DS, Bokach NA, Vasilyev AV (2016) Alkenylation of arenes and heteroarenes with alkynes. Chem Rev 116:5894–5986
82. Chatupheeraphat A, Rueping M, Magre M (2019) Chemo- and regioselective magnesium-catalyzed ortho-alkenylation of anilines. Org Lett 21:9153–9157
83. Luo J, Zhang C, Su F, Zhang B, Jia F (2020) Mechanism and origins of regio- and mono/di-selectivity in Rh(III)-Catalyzed meta-C–H alkenylation with alkynes. Eur J Org Chem 2020:3294–3302
84. Martinez AM, Echavarren J, Alonso I, Rodriguez N, Gomez Arrayas R, Carretero JC (2015) Rh(I)/Rh(III) catalyst-controlled divergent aryl/heteroaryl C–H bond functionalization of picolinamides with alkynes. Chem Sci 6:5802–5814
85. Nobushige K, Hirano K, Satoh T, Miura M (2014) Rhodium(III)-catalyzed ortho-alkenylation through C–H bond cleavage directed by sulfoxide groups. Org Lett 16:1188–1191
86. Qiu Y, Scheremetjew A, Ackermann L (2019) Electro-Oxidative C–C alkenylation by RHODIUM(III) catalysis. J Am Chem Soc 141:2731–2738
87. Qiu Z, Deng J, Zhang Z, Wu C, Li J, Liao X (2016) Mechanism of the Rhodium(III)-catalyzed alkenylation reaction of N-phenoxyacetamide with styrene or N-tosylhydrazone: a computational study. Dalton Trans 45:8118–8126

88. Ryu J, Cho SH, Chang S (2012) A versatile Rhodium(I) catalyst system for the addition of heteroarenes to both alkenes and alkynes by a C–H bond activation. Angew Chem Int Ed 51:3677–3681

89. Schipper DJ, Hutchinson M, Fagnou K (2010) Rhodium(III)-catalyzed intermolecular hydroarylation of alkynes. J Am Chem Soc 132:6910–6911

90. Shi L, Zhong X, She H, Lei Z, Li F (2015) Manganese catalyzed C–H functionalization of indoles with alkynes to synthesize bis/trisubstituted indolylalkenes and carbazoles: the acid is the key to control selectivity. Chem Commun 51:7136–7139

91. Shibata K, Natsui S, Chatani N (2017) Rhodium-catalyzed alkenylation of C–H bonds in aromatic amides with alkynes. Org Lett 19:2234–2237

92 Tan YX, Liu XY, Zhang SQ, Xie PP, Wang X, Feng KR, Yang SQ, He ZT, Hong X, Tian P, Lin GQ (2020) An unconventional trans-exo-selective cyclization of alkyne-tethered cyclohexa-dienones initiated by Rhodium(III)-catalyzed C–H activation via insertion relay. CCS Chem 2:1582–1595

93. Unoh Y, Yokoyama Y, Satoh T, Hirano K, Miura M (2016) Regioselective synthesis of benzo[b]phosphole derivatives via direct ortho-alkenylation and cyclization of arylthiophos-phinamides. Org Lett 18:5436–5439

94. Xu HJ, Kang YS, Shi H, Zhang P, Chen YK, Zhang B, Liu ZQ, Zhao J, Sun WY, Yu JQ, Lu Y (2019) Rh(III)-Catalyzed meta-C–H alkenylation with alkynes. J Am Chem Soc 141:76–79

95. Zhao H, Xu X, Luo Z, Cao L, Li B, Li H, Xu L, Fan Q, Walsh PJ (2019) Rhodium(I)-catalyzed C6-selective C–H alkenylation and polyenylation of 2-pyridones with alkenyl and conjugated polyenyl carboxylic acids. Chem Sci 10:10089–10096

96. Liu B, Zhou T, Li B, Xu S, Song H, Wang B (2014) Rhodium(III)-catalyzed alkenylation reactions of 8-methylquinolines with alkynes by C($sp3$)-H activation. Angew Chem Int Ed 53:4191–4195

97. Jiang J, Ramozzi R, Morokuma K (2015) RhIII-catalyzed C($sp3$) -H bond activation by an external base metalation/deprotonation mechanism: a theoretical study. Chem Eur J 21:11158–11164

98. Fukui Y, Liu P, Liu Q, He ZT, Wu NY, Tian P, Lin G-Q (2014) Tunable arylative cyclization of 1,6-Enynes triggered by Rhodium(III)-catalyzed C–H activation. J Am Chem Soc 136:15607–15614

99. Du L, Xu Y, Yang S, Li J, Fu X (2016) Computational insights into the Rhodium(III)-catalyzed coupling of benzamides and 1,6-enynes via a tunable arylative cyclization. J Org Chem 81:1921–1929

100. Yang YF, Houk KN, Wu YD (2016) Computational exploration of RhIII/RhV and RhIII/RhI catalysis in Rhodium(III)-catalyzed C-H activation reactions of N-phenoxyacetamides with alkynes. J Am Chem Soc 138:6861–6868

101. Han SH, Choi M, Jeong T, Sharma S, Mishra NK, Park J, Oh JS, Kim WJ, Lee JS, Kim IS (2015) Rhodium-catalyzed C–H alkylation of indolines with allylic alcohols: direct access to beta-aryl carbonyl compounds. J Org Chem 80:11092–11099

102. Liu B, Fan Y, Gao Y, Sun C, Xu C, Zhu J (2012) Rhodium(III)-catalyzed nitroso directed C–H olefination of arenes high-yield, versatile coupling under mild conditions. J Am Chem Soc 135:468–473

103. Liu B, Hu P, Zhou X, Bai D, Chang J, Li X (2017) Cp*Rh(III)-catalyzed mild addition of C(sp3)-H bonds to α, β-unsaturated aldehydes and ketones. Org Lett 19:2086–2089

104. Patureau FW, Besset T, Kuhl N, Glorius F (2011) Diverse strategies toward indenol and fulvene derivatives: Rh-catalyzed C–H activation of aryl ketones followed by coupling with internal alkynes. J Am Chem Soc 133:2154–2156

105. Takahashi H, Inagaki S, Yoshii N, Gao F, Nishihara Y, Takagi K (2009) Rh-catalyzed negishi alkyl-aryl cross-coupling leading to α- or β-phosphoryl-substituted alkylarenes. J Org Chem 74:2794–2797

106. Wang H, Yu S, Qi Z, Li X (2015) Rh(III)-catalyzed C–H alkylation of arenes using alkylboron reagents. Org Lett 17:2812–2815

107. Zhang C, Wang M, Fan Z, Sun LP, Zhang A (2014) Substituent-enabled oxidative dehydrogenative cross-coupling of 1,4-naphthoquinones with alkenes. J Org Chem 79:7626–7632

108. Brand JP, Charpentier J, Waser J (2009) Direct alkynylation of indole and pyrrole heterocycles. Angew Chem Int Ed 48:9346–9349

109. Brand JP, Chevalley C, Scopelliti R, Waser J (2012) Ethynyl benziodoxolones for the direct alkynylation of heterocycles: structural requirement, improved procedure for pyrroles, and insights into the mechanism. Chem Eur J 18:5655–5666

110. Brand JP, Chevalley C, Waser J (2011) One-pot gold-catalyzed synthesis of 3-silylethynyl indoles from unprotected o-alkynylanilines Beilstein. J Org Chem 7:565–569

111. Tolnai GL, Ganss S, Brand JP, Waser J (2013) C2-selective direct alkynylation of indoles. Org Lett 15:112–115

112. Waser J, Brand J (2012) Synthesis of 1-[(Triisopropylsilyl)ethynyl]-1λ3,2-benziodoxol-3(1H)-one and Alkynylation of Indoles, Thiophenes, and Anilines. Synthesis 44:1155–1158

113. Li Y, Xie F, Li X (2016) Formal gold- and rhodium-catalyzed regiodivergent C–H alkynylation of 2-pyridones. J Org Chem 81:715–722

114. Wang H, Xie F, Qi Z, Li X (2015) Iridium- and rhodium-catalyzed C–H activation and formyl alkynylation of benzaldehydes under chelation-assistance. Org Lett 17:920–923

115. Xie F, Qi Z, Yu S, Li X (2014) Rh(III)- and Ir(III)-catalyzed C–H alkynylation of arenes under chelation assistance. J Am Chem Soc 136:4780–4787

116. Feng C, Feng D, Luo Y, Loh T-P (2014) Rhodium(III)-catalyzed olefinic C–H alkynylation of enamides at room temperature. Chem Commun 50:9865–9868

117. Feng C, Feng D, Luo Y, Loh T-P (2014) Rhodium(III)-catalyzed olefinic C–H alkynylation of acrylamides using tosyl-imide as directing group. Org Lett 16:5956–5959

118. Feng C, Loh TP (2014) Rhodium-catalyzed C–H alkynylation of arenes at room temperature. Angew Chem 126:2760–2764

119. Collins KD, Lied F, Glorius F (2014) Preparation of conjugated 1,3-enynes by Rh(III)-catalysed alkynylation of alkenes via C-H activation. Chem Commun 50:4459–4461

120. Shaikh AC, Shinde DR, Patil NT (2016) Gold vs rhodium catalysis: tuning reactivity through catalyst control in the C–H alkynylation of isoquinolones. Org Lett 18:1056–1059

121. Zhao F, Xu B, Ren D, Han L, Yu Z, Liu T (2018) C–H Alkynylation of N-methylisoquinolone by rhodium or gold catalysis: theoretical studies on the mechanism, regioselectivity, and role of TIPS-EBX. Organometallics 37:1026–1033

122. Asaumi T, Matsuo T, Fukuyama T, Ie Y, Kakiuchi F, Chatani N (2004) Ruthenium-and rhodium-catalyzed direct carbonylation of the ortho C–H bond in the benzene ring of N-arylpyrazoles. J Org Chem 69:4433–4440

123. Guan ZH, Ren ZH, Spinella SM, Yu S, Liang Y-M, Zhang X (2009) Rhodium catalyzed direct oxidative carbonylation of aromatic C–H bond with CO and alcohols. J Am Chem Soc 131:729–733

124. Lang R, Wu J, Shi L, Xia C, Li F (2011) Regioselective Rh-catalyzed direct carbonylation of indoles to synthesize indole-3-carboxylates. Chem Commun 47:12553–12555

125. Li X, Li X, Jiao N (2015) Rh-catalyzed construction of quinolin-2(1H)-ones via C–H bond activation of simple anilines with CO and alkynes. J Am Chem Soc 137:9246–9249

126. Wang GW, Bower JF (2018) Modular access to azepines by directed carbonylative C–C bond activation of aminocyclopropanes. J Am Chem Soc 140:2743–2747

127. Yutaka Ishii NC, Kakiuchi Fumitoshi, Murai Shinji (1997) Rhodium catalyzed reaction of N-(2-Pyridinyl)piperazines with CO and ethylene. a novel carbonylation at a C–H bond in the piperazine ring. Organometallics 16:3615–3622

128. Gao B, Liu S, Lan Y, Huang H (2016) Rhodium-catalyzed cyclocarbonylation of ketimines via C–H bond activation. Organometallics 35:1480–1487

129. Cho SH, Kim JY, Kwak J, Chang S (2011) Recent advances in the transition metal-catalyzed twofold oxidative C–H bond activation strategy for C–C and C–N bond formation. Chem Soc Rev 40:5068–5083

130. Park S, Seo B, Shin S, Son JY, Lee PH (2013) Rhodium-catalyzed oxidative coupling through C–H activation and annulation directed by phosphonamide and phosphinamide groups. Chem Commun (Camb) 49:8671–8673

131. Sun CL, Li BJ, Shi ZJ (2011) Direct C–H transformation via iron catalysis. Chem Rev 111:1293–1314
132. Chen S, Yu J, Jiang Y, Chen F, Cheng J (2013) Rhodium-catalyzed direct annulation of aldehydes with alkynes leading to indenones: proceeding through in situ directing group formation and removal. Org Lett 15:4754–4757
133. Gao Y, Nie J, Li Y, Li X, Chen Q, Huo Y, Hu X-Q (2020) Rh-Catalyzed C–H amination/annulation of acrylic acids and anthranils by using -COOH as a deciduous directing group: an access to diverse quinolines. Org Lett 22:2600–2605
134. Han Z, Li P, Zhang Z, Chen C, Wang Q, Dong XQ, Zhang X (2016) Highly enantioselective synthesis of chiral succinimides via Rh/bisphosphine-thiourea-catalyzed asymmetric hydrogenation. ACS Catal 6:6214–6218
135. Sherikar MS, Prabhu KR (2019) Weak coordinating carboxylate directed Rhodium(III)-Catalyzed C–H activation: switchable decarboxylative heck-type and [4+1] annulation reactions with maleimides. Org Lett 21:4525–4530
136. Neely JM, Rovis T (2013) Rh(III)-catalyzed regioselective synthesis of pyridines from alkenes and alpha, beta-unsaturated oxime esters. J Am Chem Soc 135:66–69
137. Wang K, Hu F, Zhang Y, Wang J (2015) Directing group-assisted transition-metal-catalyzed vinylic C–H bond functionalization. Sci China Chem 58:1252–1265
138. Youn SW, Yoo HJ (2017) One-Pot sequential N-heterocyclic Carbene/Rhodium(III) catalysis: synthesis of fused polycyclic isocoumarins. Adv Synth Catal 359:2176–2183
139. Zheng L, Ju J, Bin Y, Hua R (2012) Synthesis of isoquinolines and heterocycle-fused pyridines via three-component cascade reaction of aryl ketones, hydroxylamine, and alkynes. J Org Chem 77:5794–5800
140. Hesp KD, Bergman RG, Ellman JA (2011) Expedient synthesis of N-acyl anthranilamides and beta-enamine amides by the Rh(III)-catalyzed amidation of aryl and vinyl C–H bonds with isocyanates. J Am Chem Soc 133:11430–11433
141. Piou T, Rovis T (2018) Electronic and steric tuning of a prototypical piano stool complex: Rh(III) catalysis for C–H functionalization. Acc Chem Res 51:170–180
142. Qi Z, Wang M, Li X (2013) Access to indenones by rhodium(III)-catalyzed C–H annulation of arylnitrones with internal alkynes. Org Lett 15:5440–5443
143. Wang F, Qi Z, Sun J, Zhang X, Li X (2013) Rh(III)-catalyzed coupling of benzamides with propargyl alcohols via hydroarylation-lactonization. Org Lett 15:6290–6293
144. Lei ZQ, Ye JH, Sun J, Shi ZJ (2014) Direct alkenyl C–H functionalization of cyclic enamines with carboxylic acids via Rh catalysis assisted by hydrogen bonding. Org Chem Front 1:634–638
145. Potter TJ, Kamber DN, Mercado BQ, Ellman JA (2017) Rh(III)-catalyzed aryl and alkenyl C–H bond addition to diverse nitroalkenes. ACS Catal 7:150–153
146. Webster-Gardiner MS, Chen J, Vaughan BA, McKeown BA, Schinski W, Gunnoe TB (2017) Catalytic synthesis of "super" linear alkenyl arenes using an easily prepared Rh(I) catalyst. J Am Chem Soc 139:5474–5480
147. Hawkes KJ, Cavell KJ, Yates BF (2008) Rhodium-catalyzed C–C coupling reactions: mechanistic considerations. Organometallics 27:4758–4771
148. Zhao D, Vasquez-Cespedes S, Glorius F (2015) Rhodium(III)-catalyzed cyclative capture approach to diverse 1-aminoindoline derivatives at room temperature. Angew Chem Int Ed 54:1657–1661
149. Ajitha MJ, Huang KW (2016) Mechanism and regioselectivity of Rh(III)-catalyzed intermolecular annulation of aryl-substituted diazenecarboxylates and alkenes: DFT insights. Organometallics 35:450–455
150. Tian M, Bai D, Zheng G, Chang J, Li X (2019) Rh(III)-catalyzed asymmetric synthesis of axially chiral biindolyls by merging C–H activation and nucleophilic cyclization. J Am Chem Soc 141:9527–9532
151. Yang W, Wang J, Wei Z, Zhang Q, Xu X (2016) Kinetic control of Rh(III)-catalyzed annulation of C–H bonds with quinones: chemoselective synthesis of hydrophenanthridinones and phenanthridinones. J Org Chem 81:1675–1680

152. Yi W, Li L, Chen H, Ma K, Zhong Y, Chen W, Gao H, Zhou Z (2018) Rh(III)-catalyzed oxidative [5+2] annulation using two transient assisting groups: stereospecific assembly of 3-alkenylated benzoxepine framework. Org Lett 20:6812–6816

153. Yi W, Chen W, Liu FX, Zhong Y, Wu D, Zhou Z, Gao H (2018) Rh(III)-catalyzed and solvent-controlled chemoselective synthesis of chalcone and benzofuran frameworks via synergistic dual directing groups enabled regioselective C–H functionalization: a combined experimental and computational study. ACS Catal 8:9508–9519

154. Quiñones N, Seoane A, García-Fandiño R, Mascareñas JL, Gulías M (2013) Rhodium(III)-catalyzed intramolecular annulations involving amide-directed C–H activations: synthetic scope and mechanistic studies. Chem Sci 4:2874–2879

155. Algarra AG, Cross WB, Davies DL, Khamker Q, Macgregor SA, McMullin CL, Singh K (2014) Combined experimental and computational investigations of rhodium- and ruthenium-catalyzed C–H functionalization of pyrazoles with alkynes. J Org Chem 79:1954–1970

156. Zheng J, Wang SB, Zheng C, You SL (2015) Asymmetric dearomatization of naphthols via a Rh-Catalyzed C(sp^2)-H functionalization/annulation reaction. J Am Chem Soc 137:4880–4883

157. Zheng C, Zheng J, You S-L (2016) A DFT study on Rh-catalyzed asymmetric dearomatization of 2-Naphthols Initiated with C–H activation: a refined reaction mechanism and origins of multiple selectivity. ACS Catal 6:262–271

158. Seo J, Park Y, Jeon I, Ryu T, Park S, Lee PH (2013) Synthesis of phosphaisocoumarins through rhodium-catalyzed cyclization using alkynes and arylphosphonic acid monoesters. Org Lett 15:3358–3361

159. Liu L, Wu Y, Wang T, Gao X, Zhu J, Zhao Y (2014) Mechanism, reactivity, and selectivity in Rh(III)-catalyzed phosphoryl-directed oxidative C–H activation/cyclization: a DFT study. J Org Chem 79:5074–5081

160. Fan Z, Song S, Li W, Geng K, Xu Y, Miao ZH, Zhang A (2015) Rh(III)-catalyzed redox-neutral C–H activation of pyrazolones: an economical approach for the synthesis of N-substituted indoles. Org Lett 17:310–313

161. Zhang G, Yang L, Wang Y, Xie Y, Huang H (2013) An efficient Rh/O2 catalytic system for oxidative C–H activation/annulation: evidence for Rh(I) to Rh(III) oxidation by molecular oxygen. J Am Chem Soc 135:8850–8853

162. Zhang X, Si W, Bao M, Asao N, Yamamoto Y, Jin T (2014) Rh(III)-catalyzed regioselective functionalization of C–H bonds of naphthylcarbamates for oxidative annulation with alkynes. Org Lett 16:4830–4833

163. Pham MV, Cramer N (2014) Rhodium(III)/copper(II)-promoted trans-selective heteroaryl acyloxylation of alkynes: stereodefined access to trans-enol esters. Angew Chem Int Ed 53:14575–14579

164. Zhou B, Du J, Yang Y, Feng H, Li Y (2014) Rh(III)-catalyzed C–H amidation with N-hydroxycarbamates: a new entry to N-carbamate-protected arylamines. Org Lett 16:592–595

165. Huang X, Huang J, Du C, Zhang X, Song F, You J (2013) N-oxide as a traceless oxidizing directing group: mild rhodium(III)-catalyzed C–H olefination for the synthesis of ortho-alkenylated tertiary anilines. Angew Chem Int Ed 52:12970–12974

166. Li Y, Chen H, Qu L-B, Houk KN, Lan Y (2019) Origin of regiochemical control in Rh(III)/Rh(V)-catalyzed reactions of unsaturated oximes and alkenes to form pyridines. ACS Catal 9:7154–7165

167. Liu B, Fan Y, Gao Y, Sun C, Xu C, Zhu J (2013) Rhodium(III)-catalyzed N-nitroso-directed C-H olefination of arenes. High-yield, versatile coupling under mild conditions. J Am Chem Soc 135:468–473

168. Liu G, Shen Y, Zhou Z, Lu X (2013) Rhodium(III)-catalyzed redox-neutral coupling of N-phenoxyacetamides and alkynes with tunable selectivity. Angew Chem Int Ed 52:6033–6037

169. Wangweerawong A, Bergman RG, Ellman JA (2014) Asymmetric synthesis of alpha-branched amines via Rh(III)-catalyzed C-H bond functionalization. J Am Chem Soc 136:8520–8523

170. Xie W, Chen X, Shi J, Li J, Liu R (2019) Synthesis of 1-aminoindole derivatives via Rh(III)-catalyzed annulation reactions of hydrazines with sulfoxonium ylides. Org Chem Front 6:2662–2666

171. Zhang J, Xie H, Zhu H, Zhang S, Reddy Lonka M, Zou H (2019) Chameleon-like behavior of the directing group in the Rh(III)-catalyzed regioselective C–H amidation of indole: an experimental and computational study. ACS Catal 9:10233–10244

172. Guimond N, Gorelsky SI, Fagnou K (2011) Rhodium(III)-catalyzed heterocycle synthesis using an internal oxidant: improved reactivity and mechanistic studies. J Am Chem Soc 133:6449–6457

173. Xu L, Zhu Q, Huang G, Cheng B, Xia Y (2012) Computational elucidation of the internal oxidant-controlled reaction pathways in Rh(III)-catalyzed aromatic C–H functionalization. J Org Chem 77:3017–3024

174. Neufeldt SR, Jimenez-Oses G, Huckins JR, Thiel OR, Houk KN (2015) Pyridine n-oxide vs pyridine substrates for Rh(III)-catalyzed oxidative C–H bond functionalization. J Am Chem Soc 137:9843–9854

175. Huckins JR, Bercot EA, Thiel OR, Hwang T-L, Bio MM (2013) Rh(III)-catalyzed C–H activation and double directing group strategy for the regioselective synthesis of naphthyridinones. J Am Chem Soc 135:14492–14495

176. Wu W, Liu Y, Bi S (2015) Mechanistic insight into conjugated N–N bond cleavage by Rh(III)-catalyzed redox-neutral C–H activation of pyrazolones. Org Biomol Chem 13:8251–8260

177. Chen W-J, Lin Z (2015) Rhodium(III)-catalyzed hydrazine-directed C–H activation for indole synthesis: mechanism and role of internal oxidant probed by DFT studies. Organometallics 34:309–318

178. Zhao D, Shi Z, Glorius F (2013) Indole synthesis by rhodium(III)-catalyzed hydrazine-directed C–H activation: redox-neutral and traceless by N–N bond cleavage. Angew Chem Int Ed 52:12426–12429

179. Wang Q, Li Y, Qi Z, Xie F, Lan Y, Li X (2016) Rhodium(III)-catalyzed annulation between N-sulfinyl ketoimines and activated olefins: C–H activation assisted by an oxidizing N–S bond. ACS Catal 6:1971–1980

180. Chen X-Y, Gao Z-H, Ye S (2020) Bifunctional n-heterocyclic carbenes derived from L-Pyroglutamic acid and their applications in enantioselective organocatalysis. Acc Chem Res 53:690–702

181. Chen XY, Li S, Vetica F, Kumar M, Enders D (2018) N-heterocyclic-carbene-catalyzed domino reactions via two or more activation modes. IScience 2:1–26

182. Jeletic MS, Ghiviriga I, Abboud KA, Veige AS (2007) Mono- and bimetallic rhodium(I) complexes supported by newc-symmetric bis-n-heterocyclic carbene ligands: metalation via C = C bond cleavage under Mild Conditions. Organometallics 26:5267–5270

183. Nair V, Rajesh C, Vinod AU, Bindu S, Sreekanth AR, Mathen JS, Balagopal L (2003) Strategies for heterocyclic construction via novel multicomponent reactions based on isocyanides and nucleophilic carbenes. Acc Chem Res 36:899–907

184. Frenking G, Solà M, Vyboishchikov SF (2005) Chemical bonding in transition metal carbene complexes. J Org Chem 690:6178–6204

185. Sierra MA (2000) Di- and polymetallic heteroatom stabilized (Fischer) metal carbene complexes. Chem Rev 100:3591–3638

186. Velazquez HD, Verpoort F (2012) N-Heterocyclic carbene transition metal complexes for catalysis in aqueous media. Chem Soc Rev 41:7032–7060

187. Xia Y, Zhang Y, Wang J (2013) Catalytic cascade reactions involving metal carbene migratory insertion. ACS Catal 3:2586–2598

188. Chan WW, Lo SF, Zhou Z, Yu WY (2012) Rh-catalyzed intermolecular carbenoid functionalization of aromatic C–H bonds by alpha-diazomalonates. J Am Chem Soc 134:13565–13568

189. Lu Y-S, Yu W-Y (2016) Cp*Rh(III)-catalyzed cross-coupling of alkyltrifluoroborate with alpha-diazomalonates for $C(sp^3)$-$C(sp^3)$ bond formation. Org Lett 18:1350–1353

190. Xia Y, Feng S, Liu Z, Zhang Y, Wang J (2015) Rhodium(I)-catalyzed sequential $C(sp)$-$C(sp^3)$ and $C(sp^3)$-$C(sp^3)$ bond formation through migratory carbene insertion. Angew Chem Int Ed 54:7891–7894

191. Xia Y, Liu Z, Feng S, Ye F, Zhang Y, Wang J (2015) Rh(I)-catalyzed cross-coupling of alpha-diazoesters with arylsiloxanes. Org Lett 17:956–959

192. Nakamura E, Yoshikai N, Yamanaka M (2002) Mechanism of C–H bond activation/C–C bond formation reaction between diazo compound and alkane catalyzed by dirhodium tetracarboxylate. J Am Chem Soc 124:7181–7192

193. Xie Q, Song XS, Qu D, Guo LP, Xie ZZ (2015) DFT Study on the Rhodium(II)-Catalyzed C–H Functionalization of Indoles: Enol versus Oxocarbenium Ylide. Organometallics 34:3112–3119

194. Yu S, Liu S, Lan Y, Wan B, Li X (2015) Rhodium-catalyzed C–H activation of phenacyl ammonium salts assisted by an oxidizing C–N bond: a combination of experimental and theoretical studies. J Am Chem Soc 137:1623–1631

195. Zhou T, Guo W, Xia Y (2015) Rh(V) -Nitrenoid as a key intermediate in Rh(III) –Catalyzed heterocyclization by C–H activation: a computational perspective on the cycloaddition of benzamide and diazo compounds. Chem Eur J 21:9209–9218

196. Hyster TK, Ruhl KE, Rovis T (2013) A coupling of benzamides and donor/acceptor diazo compounds to form gamma-lactams via Rh(III)-catalyzed C–H activation. J Am Chem Soc 135:5364–5367

197. Cui S, Zhang Y, Wang D, Wu Q (2013) Rh(III)-catalyzed C-H activation/[4+3] cycloaddition of benzamides and vinylcarbenoids: facile synthesis of azepinones. Chem Sci 4:3912–3916

198. Zhou B, Chen Z, Yang Y, Ai W, Tang H, Wu Y, Zhu W, Li Y (2015) Redox-neutral Rhodium-catalyzed C–H functionalization of arylamine N-oxides with diazo compounds: primary $C(sp^3)$-H/$C(sp^2)$-H activation and oxygen-atom transfer. Angew Chem Int Ed 54:12121–12126

Chapter 4
Theoretical View of Rh-Catalyzed C–H Functionalization for the Construction of C–X Bonds (X = O, N, B, Si, or Halide)

Chun-Xiang Li, Ruopeng Bai, Song Liu, Cheng-Xing Cui, and Yu Lan

The C–heteroatom bonds can be constructed by selecting particular substrates that contain heteroatoms through Rh-catalyzed C–H bond functionalization. Despite the dominance of C–C bond formation reactions, Rh-catalyzed C–X (X = N, O, halide, B, Si) bond formation reactions through C–H bond activations have also been well developed and studied in detail recently [1–14]. Given the synthetic utility of C–heteroatom bond formation and functionalization, it is not surprising that much effort has recently been put forth towards the development of new transformations that focus on Rh-catalyzed direct functionalization of C–H bonds to C–X bonds. Despite the pertinent reviews about these transformations that have been emerged, these reviews tend to focus on catalytic reactions, and mechanistic data are often speculative. In this part, we present an overview of the theoretical studies of Rh-catalyzed C–H bond activation and C–X (X = N, O, halide, B, Si) bond formation. These theoretical studies provide valuable insight into the mechanism and the origin of the regio- and stereoselectivity for these Rh-catalyzed C–H bond activation reactions.

4.1　C–O Bond Formation

The Rh-mediated C–H activation often provides a nucleophilic aryl group, which can react with an electrophilic oxygen for the construction of C–O bond. In 2018, Yang and co-workers [15] reported a Rh(III)-catalyzed C(sp^2)–H benzoxylation reaction, where a hypervalent iodine reagent is used to offer electrophilic oxygen. The kinetic isotope effect experiment ($k_H/k_D = 3.3$) indicates that the C–H bond cleavage is likely involved in the turnover-limiting step. In addition, the cyclometalated Rh(III) complex **4-1** was separated and successfully catalyzed this benzoxylation reaction, which suggests that complex **4-1** is the plausible key intermediate that participated in the catalytic cycle (Scheme 4.1).

© The Author(s), under exclusive license to Springer Nature Singapore Pte Ltd. 2021　　97
Y. Lan et al., *Computational Advances of Rh-Catalyzed C–H Functionalization*,
SpringerBriefs in Molecular Science,
https://doi.org/10.1007/978-981-16-0432-4_4

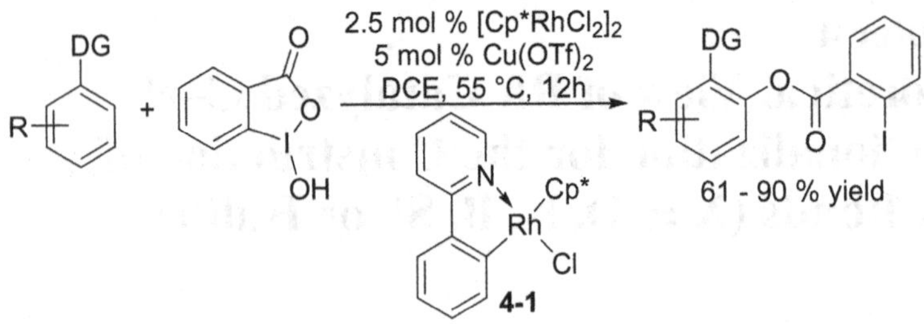

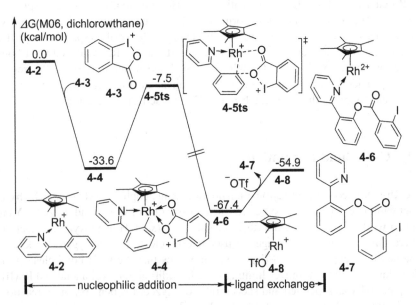

Scheme 4.1 Rh(III)-catalyzed C(sp^2)–H benzoxylation reaction using hypervalent iodine reagent as the oxygen source

The calculated free energy profiles for the C–O bond formation step are shown in Fig. 4.2 starting from five-membered rhodacyclic complex **4-2**, which can be generated from Rh-mediated C–H activation. The coordination of cationic hypervalent iodine **4-3** onto Rh(III) center in complex **4-2** produces **4-4** with 33.6 kcal/mol exothermic. In cationic hypervalent iodine, oxygen atom exhibits electrophilicity, which can react with nucleophilic aryl group on Rh(III) to achieve C–O bond formation accompanied by the cleavage of O–I bond. The activation free energy for this step is detected to be 26.1 kcal/mol via transition state **4-5ts**. Ligand exchange of **4-6** with ⁻OTf generates benzoxylation product **4-7** and cationic Rh(III) catalyst **4-8** with

Fig. 4.2 Free energy profiles for Rh(III)-catalyzed C(sp^2)–H benzoxylation reaction of 2-phenylpyridine with hypervalent iodine. The values are the relative energies given in kcal/mol calculated at the M06/6-31+G(d)/SDD//M06/6-31+G(d)/SDD level of theory in dichloroethane

12.5 kcal/mol endothermic. In this catalytic cycle, the oxidative state of Rh remains at +3 and reveals a non-redox process.

4.2 C–N Bond Formation

Nitrogen-containing compounds are widely present in natural products and synthetic intermediates. The development of efficient and selective C–N bond formation reactions has been a highly important research topic in chemical synthesis [8, 16–18]. As a result, much effort has been devoted to the conversion of an unactivated C–H bond to a C–N bond through Rh-mediated C–H bond activation in the past years [19–22]. The C–H activation can provide nucleophilic aryl-Rh or alkyl-Rh species, which can react with nitrene precursors to construct C–N bond. Alternatively, azide also can be used as a nucleophilic nitrogen source in Rh-mediated oxidative couplings.

4.2.1 Rh-Mediated C–H Amination by Using Nitrene Precursor

The Rh-nitrene complexes play important roles in C–N bond formation reactions after Rh-catalyzed C–H bond activation [23]. In this chemistry, a series of nitrene precursors can be used to construct Rh-nitrene complex, including azides [24], dioxazolones [25–27], anthranils [28–31], and N-phenoxyacetamides [32, 33]. The varied leaving group in those nitrene precursors leads to a series of different mechanisms for the generation of Rh-nitrene complexes as well as their transformations. Based on the previous experimental and theoretical studies, a common approach has been proposed in Scheme 4.3. The catalytic cycle is based on Rh-mediated C–H bond activation to form an aryl-Rh(III) intermediate 4-10. The coordination of nitrene precursor leads to the dissociation of leaving group and afford Rh-nitrene complex 4-11. Then nitrene inserts into C(aryl)–Rh(III) bond through an inner-sphere process to form C–N bond. Rh-catalyst can be regenerated by the following protonolysis. Generally, the oxidative state of Rh-nitrene complexes was deemed to +5, which is regarded as a redox process.

The azides are frequently used nitrene precursors in Rh-catalyzed arene aminations. Chang and co-workers [21, 34, 35] have made a significant contribution in Rh(III)-catalyzed intermolecular arene C–H bond aminations using organic azides as amino source. These reactions do not require external oxidants and release only nitrogen as a byproduct. Serious directing groups are effective for this reaction to furnish desired ortho-aminated products in good yields with excellent selectivity (Scheme 4.4).

They [36] also performed DFT studies to investigate the detailed mechanism of nitrene insertion, which leads to C–N bond formation. The five-membered rhodacycle

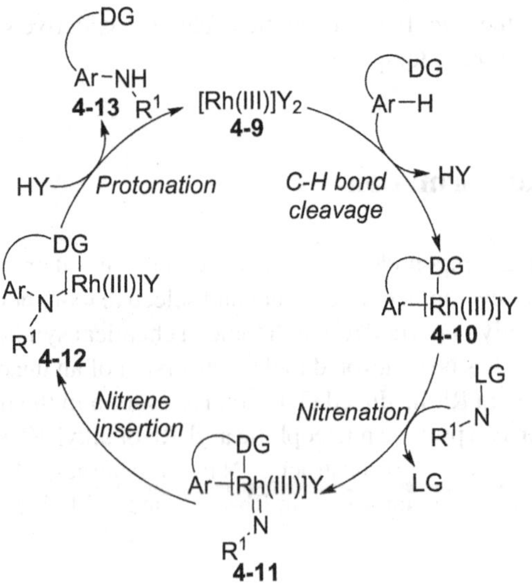

Scheme 4.3 General mechanism for Rh-mediated C–H amination by using nitrene precursor

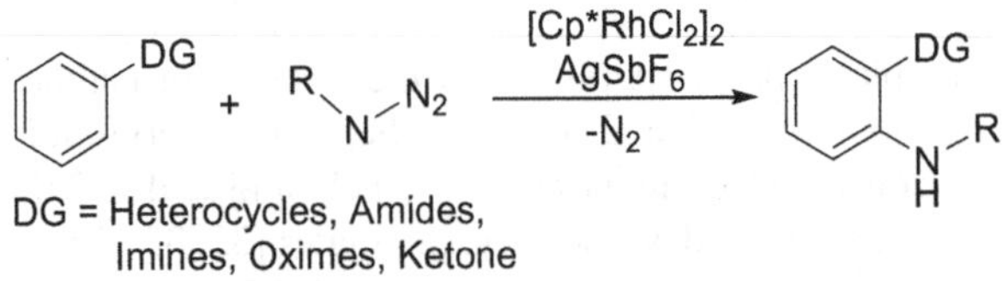

DG = Heterocycles, Amides,
Imines, Oximes, Ketone

Scheme 4.4 Rh(III)-catalyzed intermolecular C–H bond activation and amination of arenes with azides

complex **4-14**, which is separated and proved as the active intermediate in the catalytic cycle, is set as the relative zero point in the calculated free energy profiles. The ligand exchange of azide with 2-phenylpyridine in **4-14** gives the sterically matched Rh-azide species **4-16**. The subsequent denitrogenation occurs via transition state **4-17ts**, which is calculated to be the rate-determining step in the catalytic cycle, with an overall activation free energy of 28.7 kcal/mol to give the Rh(V)-nitrene complex **4-18**. Then nitrene reductive insertion into the C(aryl)–Rh bond takes place via a three-centered transition state **4-19ts** with a low energy barrier of 7.6 kcal/mol. The generated six-membered Rh(III)-amino species **4-20** was considered as a Brønsted base, which can undergo concerted metalation-deprotonation with 2-phenylpyridine via transition state **4-22ts** with an energy barrier of 24.6 kcal/mol to get the amination product **4-23** and regenerate the Rh(III) active species **4-14** by coordination of a 2-phenylpyridine substrate (Fig. 4.5).

In 2015, Chang and co-workers [26] described a robust direct Rh(III)-catalyzed C–H bond activation and amidation reaction using dioxazolones as a novel type

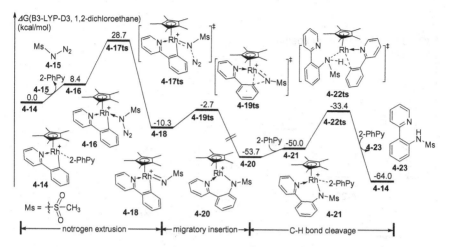

Fig. 4.5 Free energy profiles for Rh(III)-catalyzed intermolecular C–H bond activation and amination of arenes with azides. The values are the relative energies given in kcal/mol calculated at the B3-LYP-D3/6-31G(d)/SRSC-ECP//B3-LYP/6-31G(d)/SRSC-ECP level of theory in 1,2-dichloroethane

of nitrene precursors to construct C(aryl)-N(amido) bonds. In this chemistry, dioxazolones are more convenient to prepare, store, and use compared to the corresponding azides, which can release gaseous carbon dioxide to formally form an acylnitrene. Moreover, this new strategy was found to be highly efficient over a broad range of substrates with high functional group tolerance releasing carbon dioxide as a single byproduct (Scheme 4.6).

DFT calculations were taken to compare the reactivity of azide and dioxazolone in nitrenation step. As shown in Fig. 4.7, a five-membered rhodacycle **4-24** can be formed through pyridyl-directed C–H bond activation, which is set to relative zero. Dioxazolone can be activated by the coordination onto Rh in complex **4-26**, which leads to decomposition with the release of gaseous carbon dioxide. In Rh-assisted dioxazolone decomposition transition state **4-27ts**, the breaking N–O and O–C bonds are 2.49 and 1.63 angstrom, respectively, which reveal a concerted process. The calculated energy barrier for this step is only 15.1 kcal/mol. As a contrast, the

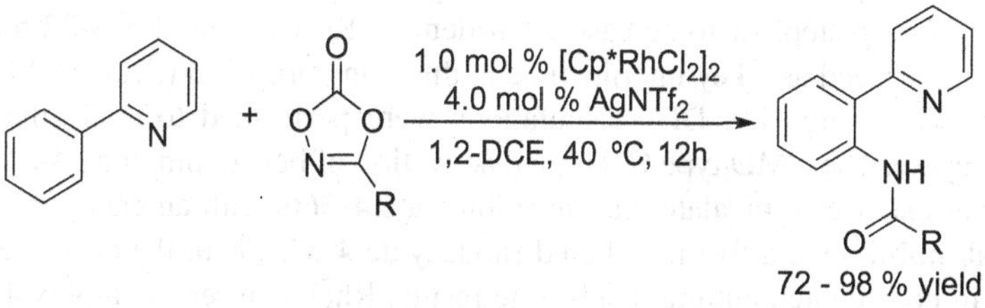

Scheme 4.6 Rh(III)-catalyzed C–H bond activation and amidation reaction of 2-phenylpyridine with 1,4,2-dioxazol-5-ones

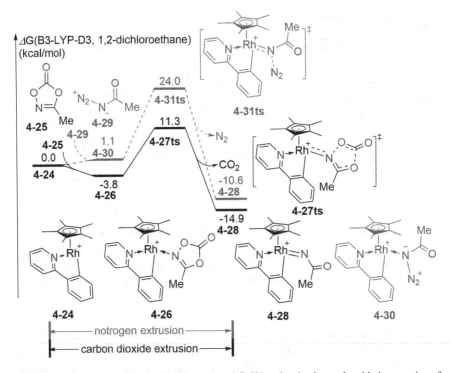

Fig. 4.7 Free energy profiles for Rh(III)-catalyzed C–H bond activation and amidation reaction of 2-phenylpyridine with 1,4,2-dioxazol-5-ones. The values are the relative energies given in kcal/mol calculated at the B3-LYP-D3/6-31G(d)/SRSC-ECP//B3-LYP/6-31G(d)/SRSC-ECP level of theory in 1,2-dichloroethane

corresponding Rh-assisted decomposition of acetyl azide can occur via transition state **4-31ts** with an energy barrier of 24.0 kcal/mol. Therefore, dioxazolone provides more reactivity in nitrenation of Rh.

The reductive cleavage of N–O bond in anthranil can provide a strong aromatic benzene ring, which was considered as the driving force for the formation of nitrene. In 2016, Li and Lan [31] developed a Rh(III)-catalyzed amination reaction of both C(aryl)–H and C(alkyl)–H bonds using anthranils as nitrene precursors. The large KIE value ($k_H/k_D = 5.3$) indicated that cleavage of the C–H bond is involved in the turnover-limiting step. In some cases, a tridentate Rh(III) complex **4-32** has been isolated and proved as a key intermediate in this transformation (Scheme 4.8).

As shown in Fig. 4.9, DFT calculations were performed to study the whole catalytic cycle. The CMD type C–H bond activation of benzoquinolone **4-34** occurs with the assistance of pivalate via transition state **4-36ts** with an energy barrier of 14.8 kcal/mol to give a five-membered rhodacycle **4-37**. Then, the cleavage of N–O bond in coordinated anthranil substrate results Rh(V)–nitrene complex **4-41** via transition state **4-40ts** with an energy barrier of 16.6 kcal/mol. The subsequent reductive insertion of nitrene into C(aryl)–Rh bond, which is the rate-determining step in

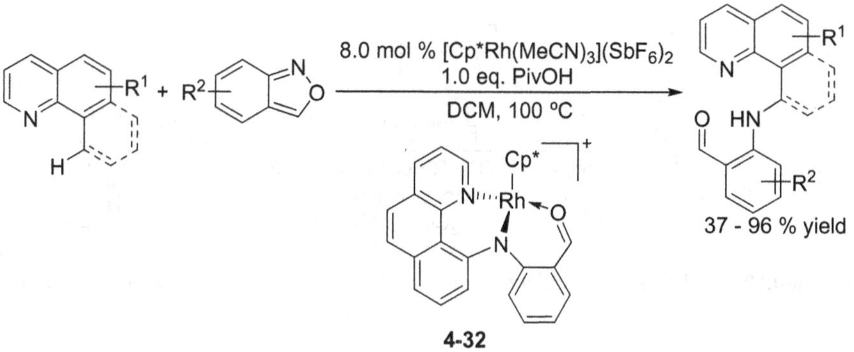

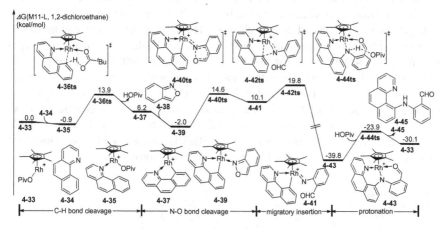

Scheme 4.8 Rh(III)-catalyzed C–H bond activation and amination reaction of 8-methylquinoline with anthranils

Fig. 4.9 Free energy profiles for Rh(III)-catalyzed C–H bond activation and amination reaction of 8-methylquinoline with anthranils. The values are the relative energies given in kcal/mol calculated at the M11-L/6-311+G(d)/LANL08(f)//B3-LYP/6-31G(d)/LANL08(f) level of theory in 1,2-dichloroethane

the catalytic cycle, takes place via transition state **4-42ts** with an overall activation free energy of 21.8 kcal/mol to form amino-Rh(III) intermediate **4-43** irreversibly. The protonation of **4-43** with pivalic acid yields the amination product **4-45** and regenerates the active catalyst **4-43** to complete the catalytic cycle.

N–O covalent bond provides an internal oxidant, which can be used to generate metal–nitrene complex. In 2017, Glorius and co-workers [37] developed a Cp*Rh(III) and oxanorbornadiene co-catalyzed C–H bond amidation reaction using intramolecular amide transfer strategy. The amide group in *N*-phenoxyacetamide acts not only as a cleavable directing group but also as an essential coupling partner for the C–H amidation (Scheme 4.10).

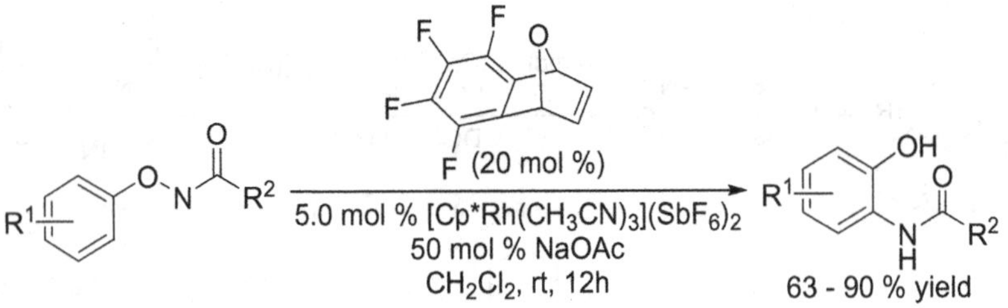

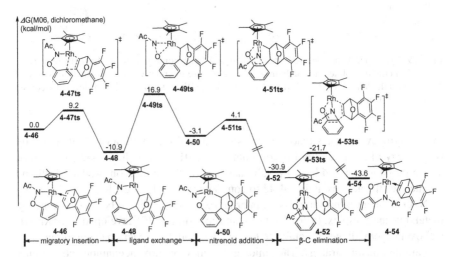

Scheme 4.10 Rh(III) and bicyclic olefin co-catalyzed C–H bond activation and amidation reaction

They also performed theoretical studies to understand the mechanism of the above reaction. [37] As shown in Fig. 4.11, the oxanorbornadiene-coordinated rhodacycle complex **4-46**, which is generated from amido directed C–H activation, was set as the relative zero point in the calculated catalytic cycle. The migratory insertion of the *exo*-face of oxanorbornadiene with a low barrier of 9.2 kcal/mol via transition state **4-47ts** gives seven-membered rhodacycle **4-48**. The subsequent oxidative elimination of phenoxyl in *N*-phenoxyacetamide by Rh proceeds via a three-membered ring-type transition state **4-49ts** to generate Rh(V)-nitrene complex **4-50**, which is considered to be the rate-determining step in the catalytic cycle with an overall activation free energy of 27.8 kcal/mol. In complex **4-50**, the electron-rich phenoxyl plays as a nucleophile, which can intramolecularly attack nitrene moiety to construct a new C–N bond in spirobicyclic intermediate **4-52**. Then, the β-C elimination of **4-52** restores aromaticity and generates the amino-Rh(III) complex **4-54**, from which the final amination product is yielded by the two steps of protonolysis.

Fig. 4.11 Free energy profiles for Cp*Rh(III) and bicyclic olefin co-catalyzed C–H bond activation and amidation reaction. The values are the relative energies given in kcal/mol calculated at the M06/6-311+G(d,p)/SDD//M06/6-31+G(d)/LANL2DZ level of theory in dichloromethane

Similar to Rh-carbene, the Rh-nitrene also can react with a C–H bond by an outer-sphere nitrene insertion into C–H bond to construct C–N bond. In 2017, Itoh and co-workers developed the first strategy of Rh(III)-catalyzed intermolecular C(sp^3)–H amination through intramolecular ligand-to-nitrene single electron transfer. The active Rh(III) catalyst **4-55** can successfully achieve the C–H amination by employing xanthene as substrate and tosyl azide (TsN$_3$) as nitrene source. KIE experiment ($k_H/k_D = 5.0$) indicated that the C(sp^3)–H bond cleavage is involved in the rate-determining step (Scheme 4.12).

The DFT calculation indicated that, in the open-shell singlet diradical state of Rh(III)-nitrene complex **4-56** and corresponding triplet diradical state, one single electron locates at nitrene moiety and another one delocalizes over the redox-active tetradentate ligand. Those two diradicals are close to each other in energy but energetically more stable than the corresponding closed-shell singlet state. The calculated free energy profile for the stepwise pathway of the C(sp^3)–H amination of xanthene is shown in Fig. 4.13. The intermolecular hydrogen abstraction from xanthene by the nitrene radical in complex **4-56** takes place via transition state **4-58ts** with a barrier of 10.8 kcal/mol to give amino-Rh(III) complex **4-60** and xanthene radical **4-59** with 20.1 kcal/mol exergonic. Then a radical rebound process occurs to yield the product-coordinated Rh(III) complex **4-62** via the singlet transition state **4-61ts** with a low barrier of only 1.1 kcal/mol. The calculated results revealed that the rate-determining step in the catalytic cycle is the C(sp^3)–H bond activation, which is consistent with KIE experiments.

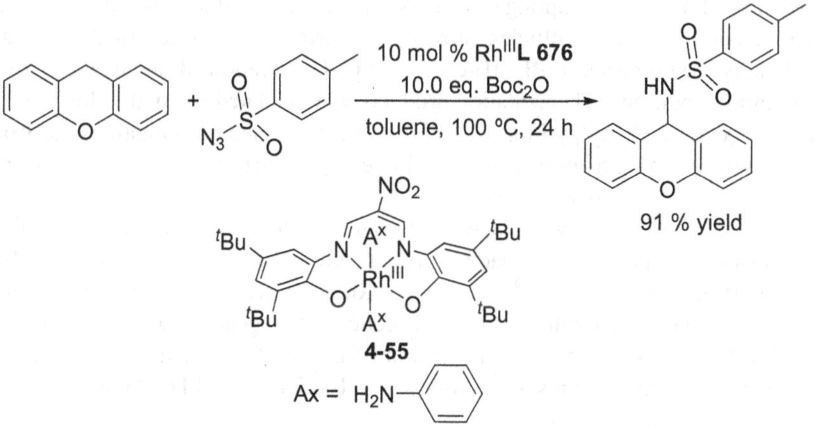

Scheme 4.12 Rh(III)-catalyzed intermolecular C(sp^3)–H activation and amination reaction of xanthene with tosyl azide

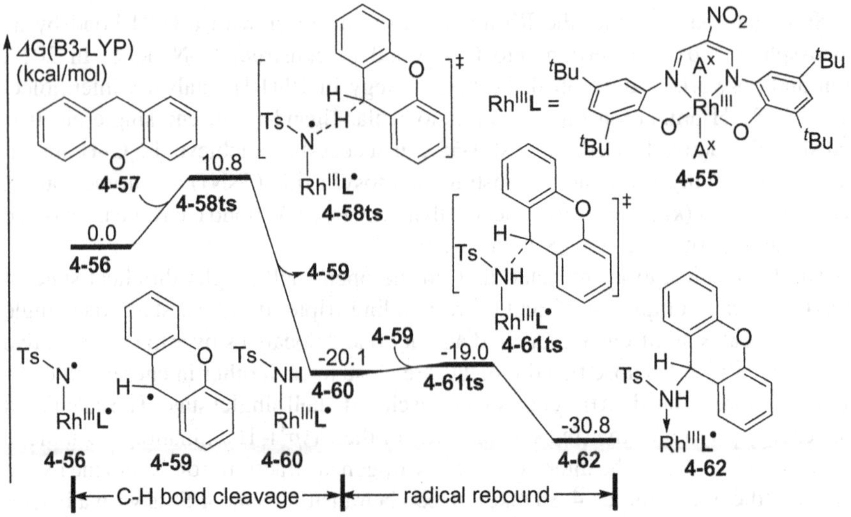

Fig. 4.13 Free energy profiles for Rh(III)-catalyzed intermolecular C(sp^3)–H activation and amination reaction of xanthene with tosyl azide. The values are the relative energies given in kcal/mol calculated at the B3-LYP/D95(d,p)/SDD level of theory

4.2.2 C–H Bond Azidation Reaction

Rh-catalyzed oxidative coupling of C–H bond with azide salts in the presence of an external oxidant is another efficient strategy to construct C–N bonds. In 2013, Li and co-workers [38] reported a Rh(III)-catalyzed C–H azidation of arenes under relatively mild conditions. The azidation products were isolated in good to high yields for a range of 2-phenylpyridines (2-PhPys) bearing electron-donating, electron-withdrawing, or halogen groups on the pyridine ring in the presence of strong external oxidant PhI(OAc)$_2$ (Scheme 4.14).

Lan and co-workers performed DFT calculation to investigate the detailed mechanism for this azidation reaction. As shown in Fig. 4.15, the active catalyst Rh(III)Cp*(OAc)Cl **4-63** combines with the oxidant PhI(OAc)OTs, which is formed by the reaction between PhI(OH)OTs and acetic acid, to generate a chloride-bridged complex **4-65**. Subsequently, the migration of acetate from trivalent iodine to Rh center occurs via transition state **4-66ts** with a barrier of 19.4 kcal/mol to give the

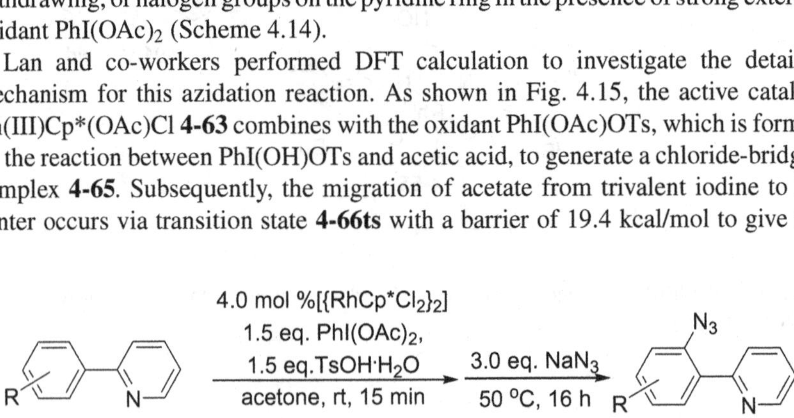

Scheme 4.14 Rh(III)-catalyzed C–H bond activation and azidation of 2-phenylpyridines with NaN$_3$

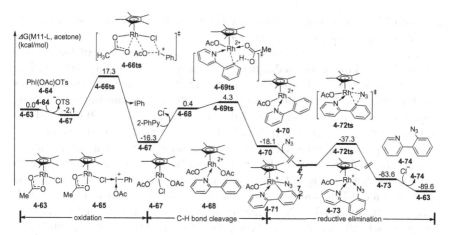

Fig. 4.15 Free energy profiles for Rh(III)-catalyzed C–H bond activation and azidation of 2-phenylpyridines with NaN$_3$. The values are the relative energies given in kcal/mol calculated at the M11-L/6-311+G(d)/SDD//B3-LYP/6-31G(d)/SDD level of theory in acetone

Rh(V) complex **4-67**. After coordination of 2-PhPy and dissociation of Cl$^-$, the acetate-assisted CMD type C–H bond cleavage of 2-PhPy takes place via transition state **4-69ts** with an overall activation free energy of 20.6 kcal/mol to give five-membered rhodacycle **4-70**. After the coordination of anionic azide, the C(aryl)-N(azide) bond coupling irreversibly gives the N–N chelated Rh(III) complex **4-73** via a three-membered ring reductive elimination transition state **4-72ts** with a barrier of only 9.9 kcal/mol. The major product 2-(2-azidophenyl)-pyridine dissociates from the catalytic cycle by ligand exchange with Cl$^-$, and the active catalyst **4-63** is regenerated to complete the catalytic cycle. In a competitive reductive elimination, azide provides better reactivity than other present nucleophiles, such as chloride and acetate.

4.3 C–Halide Bond Formation

In recent years, transition metal-catalyzed C–halogen bond formation utilizing direct functionalization of C–H bonds has consequently received considerable interest from numerous researchers, which would provide useful aryl halide for further transformations [39–46]. The vast majority of the examples of these C–halogen bond formation reactions use a palladium catalyst, while the relating works of using Rh catalysts are relatively rare [46–49]. For Rh-catalyzed C–H bond activation and C–halogen bond formation reactions, one of the best strategies is that using NXS (X = Cl, Br, I) as an electrophilic halogen source to cross-coupled with C–H bond [46, 47]. Previous theoretical studies strongly suggested that these transformations usually favor a non-redox mechanism involving X transfer pathway rather than the alternative oxidation addition of NXS following by reductive elimination through a Rh(V) species, when aryl-Rh(III) species is formed (Scheme 4.16).

In 2012, Glorius and co-workers developed the first example of the Rh(III)-catalyzed *ortho* bromination/iodination of arenes with high yielding and versatility [47]. This strategy is compatible with various highly useful directing groups. In this transformation, the NBS or NIS was used as the efficient brominating or iodinating reagent respectively. In addition, the KIE experiment ($k_H/k_D = 2.0$) indicated that the C–H bond cleavage is likely involved in the rate-determining step (Scheme 4.17).

Lan group [50] performed DFT calculations to study the mechanism and evaluate the feasibility of the formation of Rh(V) species in this Rh(III)-catalyzed bromination reaction. As shown in Fig. 4.18, the amide-directed C–H bond cleavage occurs via transition state **4-78ts** with an energy barrier of 24.6 kcal/mol to form the aryl-Rh(III) intermediate **4-79**. The following intermolecular Br transfer from NBS to the aryl moiety affords bromonium complex **4-81** via transition state **4-80ts** with a small barrier of 6.0 kcal/mol. The Br shift from C2 to C1 occurs with cleavage of C(aryl)–Rh bond via transition-state **4-82ts** with an overall activation free energy of 25.7 kcal/mol to give the bromination product coordinated Rh(III) complex **4-83**. Releasing of the bromination product **4-84** and protonation would regenerate active catalyst **4-75** to complete the catalytic cycle. In this pathway, the oxidative state of Rh remains at +3 resulting a non-redox process. Alternatively, the oxidation addition-reductive elimination pathway was ruled out due to an unfavorable activation energy of 29.1 kcal/mol. However, DFT calculations found that the Rh(III)/Rh(V) catalytic cycle, which involves a Rh(V) intermediate generated by the oxidative addition of

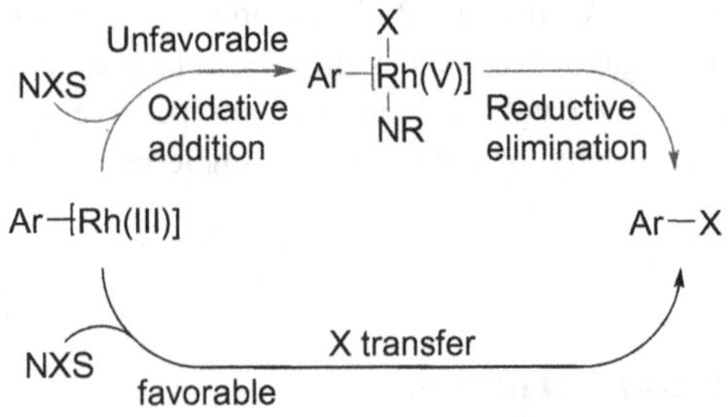

Scheme 4.16 Mechanism for Rh-catalyzed C–H bond activation and C–halogen bond formation reactions

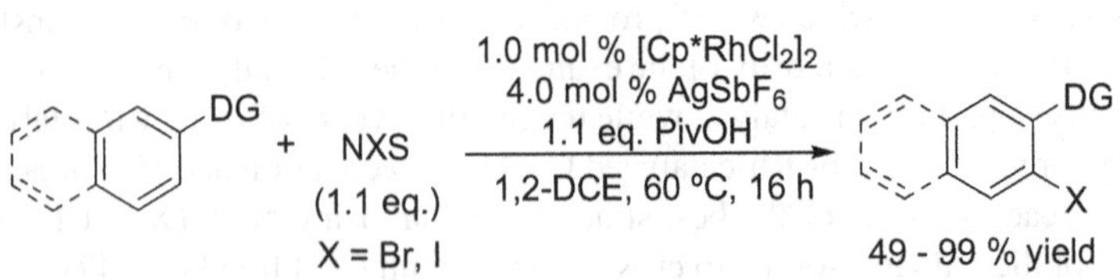

Scheme 4.17 Rh(III)-catalyzed C–H bond activation and *ortho* bromination or iodination of arenes with NBS or NIS

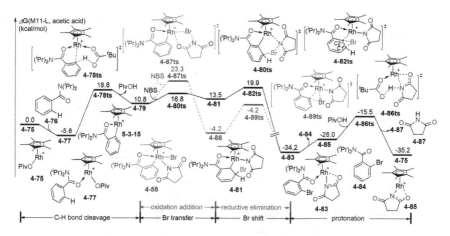

Fig. 4.18 Free energy profiles for Rh(III)-catalyzed C–H bond activation and *ortho* bromination arenes of with NBS. The values are the relative energies given in kcal/mol calculated at the M11-L/6-311+G(d)/LANL08(f)//B3-LYP/6-31+G(d)/SDD level of theory in acetic acid

NBS, is become favorable when a strong electron-withdrawing group is installed on arene (Scheme 4.19).

Mück-Lichtenfeld and Glorius also performed DFT calculations to shed light on the possible pathways of this reaction [46]. The theoretical studies indicated that the stepwise oxidation addition and reductive elimination pathway are unfavorable due to the high activation free energy of 27.3 kcal/mol through the formation of a Rh(V) complex **4-94**. Alternatively, when an aryl-Rh(III) species **4-90** is formed, an outer-sphere electrophilic chlorine transfer from NCS to aryl group takes place via a

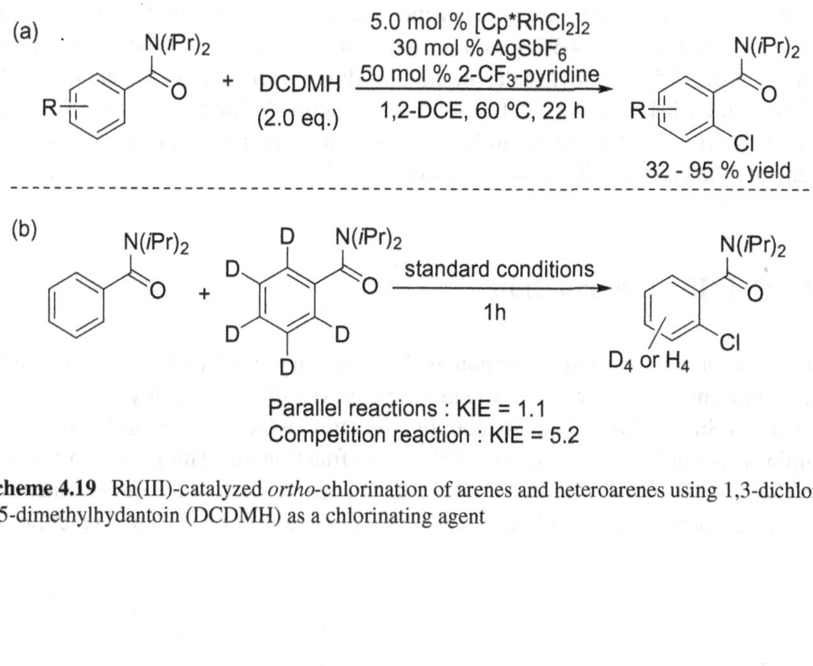

Scheme 4.19 Rh(III)-catalyzed *ortho*-chlorination of arenes and heteroarenes using 1,3-dichloro-5,5-dimethylhydantoin (DCDMH) as a chlorinating agent

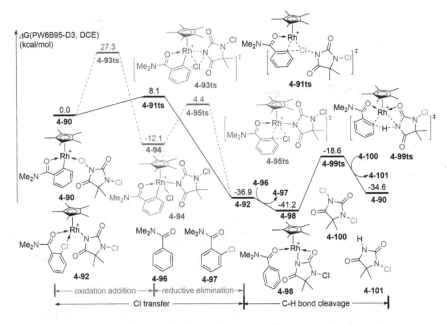

Fig. 4.20 Free energy profiles for Rh(III)-catalyzed *ortho*-chlorination of arenes and heteroarenes using 1,3-dichloro-5,5-dimethylhydantoin (DCDMH) as a chlorinating agent. The values are the relative energies given in kcal/mol calculated at the PW6B95-D3/def2-TZVP//TPSS-D3/def2-TZVP level of theory in 1, 2-dichloroethane

linear transition state **4-91ts** with an energy barrier of only 8.1 kcal/mol generating a new C(aryl)Cl bond in Rh(III) intermediate **4-92**, which is detected to be favorable. After replacement of the halogenated arene product **4-97** with benzamide **4-96** in complex **4-92**, the σ-complex assisted metathesis type C–H bond cleavage takes place via transition state **4-99ts** with a barrier of 22.6 kcal/mol. After replacement of the 1-chloro-5,5-dimethylimidazolidine-2,4-dione with the next reagent molecule DCDMH, the Rh(III) active catalyst **4-90** is regenerated. The authors proposed that the rate-determining step seems to be the dissociation of a ligand from the metal to regenerate the catalytically active species (Fig. 4.20).

4.4   C–B Bond Formation

In recent years, organoboron compounds have drawn considerable attention due to their important significance to the organic synthesis reaction especially their utility in the famous Suzuki-Miyaura cross-coupling reaction, which can be easily converted to amines, alcohols, alkenes, and other classes of functionalized molecules in a single step [51–53]. Therefore, extensive efforts have been devoted to the development of borylation reactions, especially the transition metal-catalyzed borylation reaction, in recent years [54–60]. The Rh-catalyzed C–H bond borylation reactions have attracted much attention due to high activity and high selectivity [61–63].

In a general view of the mechanism, the Rh-catalyzed borylation can undergo two steps of oxidative addition with C–H/B–X bonds and two steps of reductive elimination to form C–B/H–X bonds. Therefore, both Rh(I)/Rh(III) and Rh(III)/Rh(V) catalytic cycles are proposed depending on the order of oxidative addition and reductive elimination. As shown in Scheme 4.21a, when Rh(I)-hydride **4-102** is used as catalyst, an oxidative addition type C–H activation of reacting alkane provides Rh(III)-dihydride species **4-103**. Then reductive elimination of two hydrides results in an alkyl Rh(I) intermediate **4-104** by releasing gaseous dihydrogen, which can undergo another oxidative addition with H-BR$_2$ to form a boryl alkyl Rh(III)-hydride **4-105**. The subsequent reductive elimination yields borylation product and regenerates Rh(I)-hydride species **4-102**. In an alternative process (Scheme 4.21b), diboryl Rh(III) species **4-107** can undergo an oxidative addition with C(alkyl)–H bond

Scheme 4.21 General mechanism for Rh-catalyzed C–H bond borylation reaction

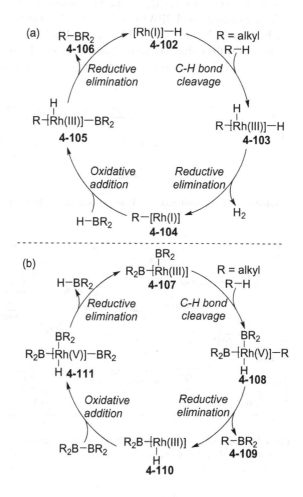

to afford alkyl-Rh(V) intermediate **4-108**, which can undergo a B(boryl)–C(alkyl) reductive elimination to yield borylation product **4-109**. The generated boryl Rh(III)–hydride **4-110** can react with diboryl through sequential B(boryl)–B(boryl) oxidative addition and B(boryl)–H reductive elimination to regenerate active species **4-107**.

When H-Bpin is used as boryl source, Rh-catalyzed C–H bond borylation reaction can undertake a dehydrogenative coupling with C–H bond. These transformations do not require external oxidants and keep redox neutral through the release of gaseous dihydrogenation. In 2001, Marder and co-workers [64] developed a highly selective Rh(I)-catalyzed borylation of C(benzyl)–H bonds with H-Bpin. The high selectivity for the benzylic C–H functionalization was observed with toluene, p-xylene, and mesitylene. The Rh(III) complex **4-112**, which is generated by oxidative addition of H-Bpin onto Rh(I) catalyst, was separated and determined by X-ray diffraction. In addition, Rh(III) complex **4-112** can also catalyze this borylation reaction, so it was considered as the active precursor in the catalytic cycle (Scheme 4.22).

Lin and Marder investigated the mechanism of benzylic C–H borylation using DFT calculations at B3PW91 level [65]. As shown in Fig. 4.23, the calculated catalytic cycle begins with a 14-electron Rh(I)-H complex **4-113**, which can be generated by the reductive elimination of Cl-Bpin form Rh(III) complex **4-112**. The benzylic C–H bond cleavage occurs via an oxidation addition type transition state

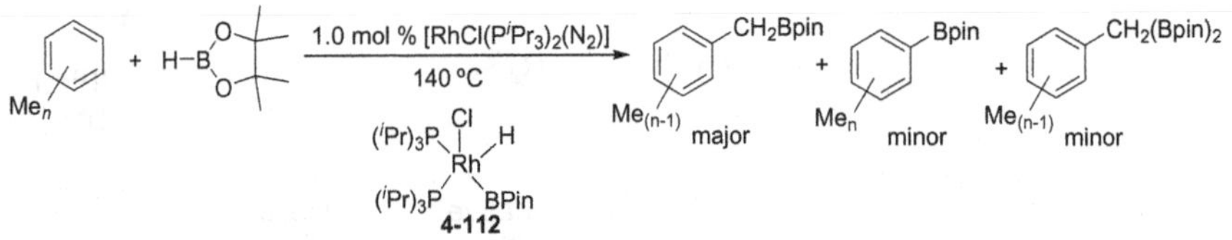

Scheme 4.22 Rh(I)-catalyzed borylation of benzylic C–H bonds with H-Bpin

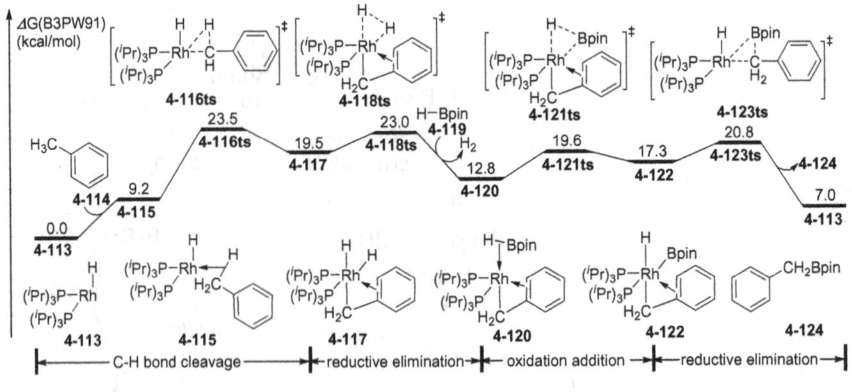

Fig. 4.23 Free energy profiles for Rh(I)-catalyzed borylation of benzylic C–H bonds with H-Bpin. The values are the relative energies given in kcal/mol calculated at the B3PW91/6-31G/LANL2DZ level of theory

4-116ts to give a η^3-benzylic Rh(III)-hydride intermediate **4-117** with a barrier of 23.5 kcal/mol. This C–H bond cleavage is considered to be the rate-determining step in the catalytic cycle. The reductive elimination of two hydrides in **4-117** and sequential substitution of dihydrogen by H-Bpin give intermediate **4-120**. The following oxidation addition of the B–H bond onto Rh(I) center takes place via transition state **4-121ts** with an energy barrier of 6.8 kcal/mol resulting in the boryl–Rh(III) complex **4-122**. Then, the reductive elimination of B-C(benzyl) bond would yield the borylated product **4-124** and regenerate the active Rh(I)-hydride active catalyst **4-113**.

In 2000, Hartwig and co-workers [60] developed a strategy of Rh-catalyzed directed borylation of alkanes to terminal alkylboronate esters with highly regioselectivity. The active bisboryl-Rh(V) species **4-125** and triboryl-Rh(V) species **4-126**, which can be observed directly in catalytic reactions, were synthesized, separated, and characterized by the same group in 2005 [66]. These compounds also can react with alkanes and arenes to form alkyl- and arylboronate esters (Scheme 4.24).

Hall and co-workers performed a DFT calculation to investigate the detailed mechanism on the model reaction of methane with the truncated model complex $HBO_2C_2H_4$ [67]. As shown in Fig. 4.25, the bisboryl-Rh(III) complex **4-127**, which is generated by reductive dissociation of borane from triboryl-Rh(V) species **4-126**, is set as the zero point. The C–H bond activation of the methane occurs via oxidation addition type transition state **4-130ts** to give the methyl-Rh(V) intermediate **4-131** with an energy barrier of 17.8 kcal/mol. Intermediate **4-131** undergoes C(methyl)–B bond formation via the reductive elimination transition state **4-132ts** with an overall activation free energy of 21.1 kcal/mol to yield borylated product **4-133** and give a 16-electron Rh(III)-hydride **4-134**. The reaction of the 16-electron fragments with diboron reagent would regenerate triboryl-Rh(V) species **4-126** to complete the catalytic cycle.

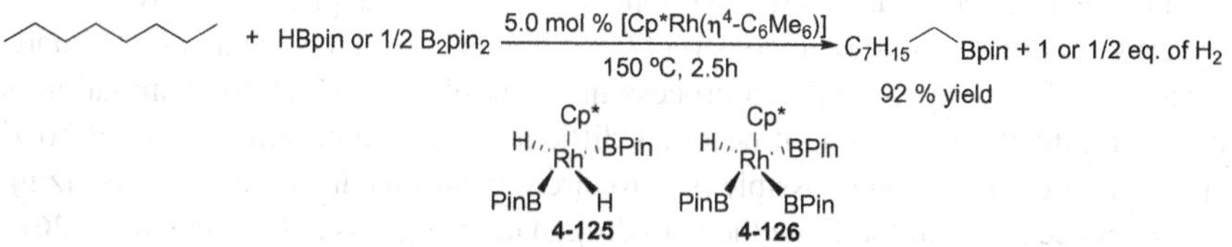

Scheme 4.24 Rh(I)-catalyzed regioselective and terminal borylation of alkanes with H-Bpin

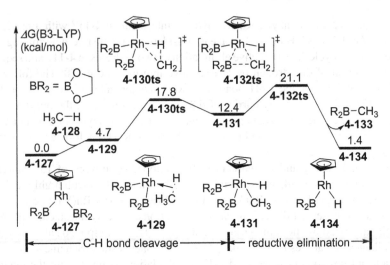

Fig. 4.25 Free energy profiles for Rh(I)-catalyzed regioselective and terminal borylation of alkanes with H-Bpin. The values are the relative energies given in kcal/mol calculated at the B3-LYP/cc-pVDZ/LANL2DZ level of theory

4.5 C–Si Bond Formation

Organosilicone compounds have recently received much attention because they are also useful precursors, similar to organoboron derivatives, in organic synthetic reactions in recent years [68–77]. The formation of C–Si bonds by transition metal-catalyzed C–H silylation is a valuable route toward the organosilicon products [78–98]. The transition metal rhodium, because of its unique catalytic activity, is usually used in the C–H bond activation and silylation reaction [99–101]. Similar to C–H bond borylations, Rh-catalyzed C–H bond silylation also applies the dehydrogenative cross-coupling strategy to construct C–Si bonds between Si–H and C–H bonds [102–109]. The dehydrogenation process in intramolecular C–H bond silylation is usually relatively easy, while it becomes difficult for the intermolecular C–H bond silylation [110–112]. One possible way to circumvent this drawback is by utilizing hydrogen acceptors to facilitate the dehydrogenation process [113] (Scheme 4.26).

In 2010, Takai and co-workers [114] developed a Rh(I)-catalyzed intramolecular dehydrogenative cross-coupling reactions between C–H bond and Si–H bond to synthesis silole derivatives in the absence of external oxidants, which produces only H_2 as a side product. The KIE of $k_H/k_D = 6.8$ highly indicated that the C–H bond

Scheme 4.26 Rh-catalyzed
C–H bond silylation applies
the dehydrogenative
cross-coupling strategy

$$R-H \ + \ [Si]-H \ \xrightarrow{\ [Rh]\ } \ H_2 \ + \ R-[Si]$$

$$+ =\!\!=$$

$$\cdots\cdots\rightarrow H_3C-CH_3$$

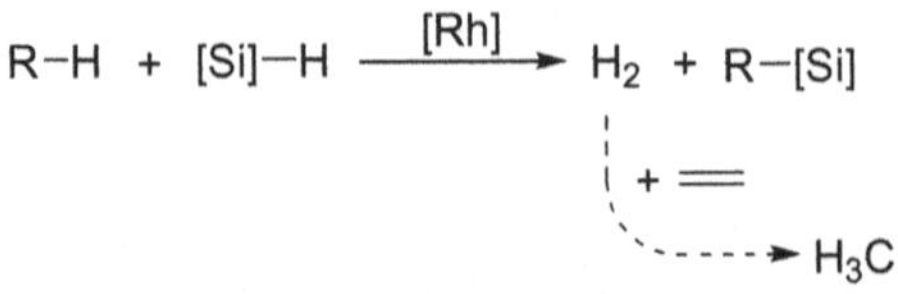

activation of the phenyl group is the rate-determining step in the catalytic cycle (Scheme 4.27).

The proposed mechanism for this intramolecular C–H bond dehydrogenative silylation is shown in Scheme 4.28. The oxidation addition of H–Si onto Rh(I) center gives the silyl-Rh(III) hydride intermediate **4-137**, which can undergo a Rh–H and C(aryl)–H bond metathesis to achieve rhodasilinane **4-139**. The following reductive elimination yields silole product **4-140** and regenerates active Rh(I) species **4-135**. The transformation from silyl-Rh(III) hydride intermediate **4-137** to rhodasilinane **4-139** also can undergo sequential oxidative addition of C(aryl)–H bond and bishydride reductive elimination through Rh(V) intermediate **4-138**.

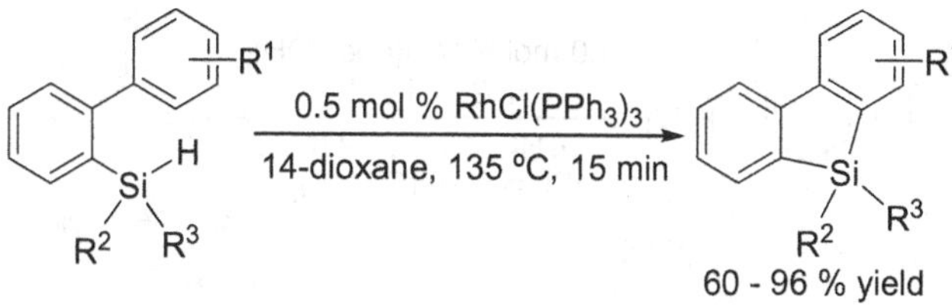

Scheme 4.27 Rh(I)-catalyzed Si–H bond and C–H bond activation

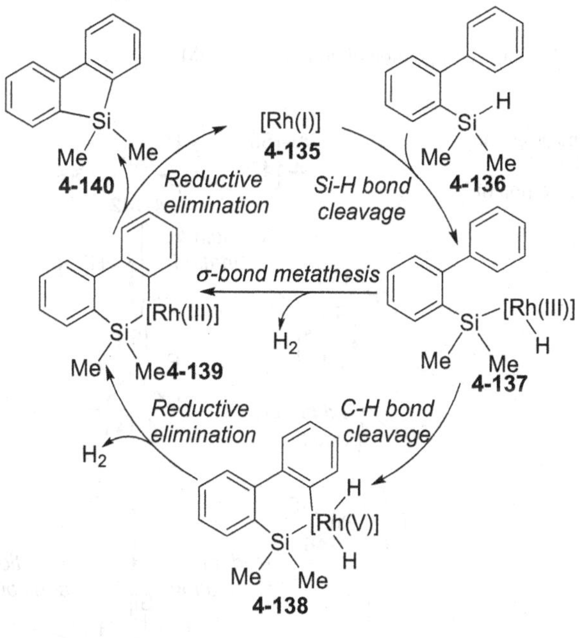

Scheme 4.28 The proposed mechanism for the Rh(I)-catalyzed Si–H bond and C–H bond activation to synthesis the silafluorene

Olefin can act as a hydrogen acceptor, which would significantly promote inter-molecular transfer dehydrogenative cross-coupling of C(aryl)–H and Si–H bonds. In 2014, Hartwig and co-workers [115] reported an Rh(I)-catalyzed intermolecular silylation of arenes with HSiMe(OSiMe$_3$)$_2$ (Scheme 4.29), which runs well under mild conditions with a hydrosilane as silicon source and cyclohexene as a hydrogen acceptor. The high regioselectivity of this silylation reaction is derived from the steric properties of substituents on the substrates and the ligands.

Based on the experimental investigations, Hartwig proposed a redox Rh(I)-Rh(III) mechanism for this transfer dehydrogenative silylation reaction (Scheme 4.30). The active catalyst in the catalytic cycle is considered to be a silyl-Rh(III) dihydride

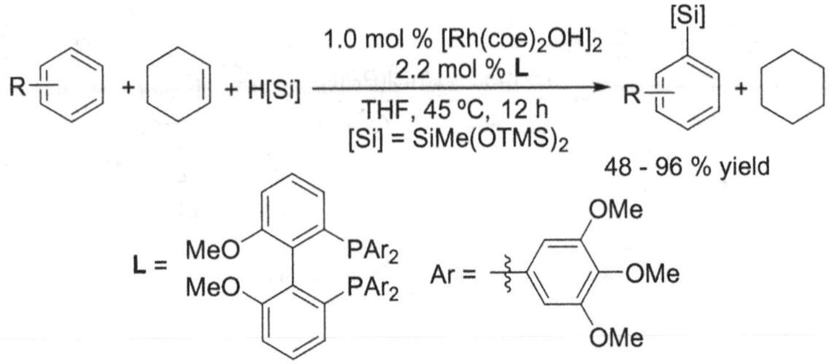

Scheme 4.29 Rh(I)-catalyzed silylation of arenes with HSiMe(OSiMe$_3$)$_2$

Scheme 4.30 The proposed mechanism for the Rh(I)-catalyzed silylation of arenes with HSiMe(OSiMe$_3$)$_2$

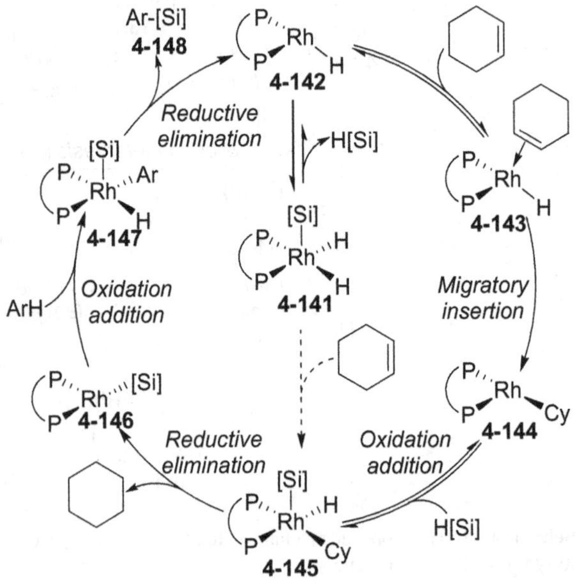

complex **4-141**, which is isolated and characterized by NMR spectroscopy and X-ray diffraction. In one possible pathway, intermolecular insertion of cyclohexene into the Rh–H bond generates cyclohexyl-silyl-Rh(III) hydride **4-145**. Alternatively, the reversible reductive elimination of Si-H in **4-141** forms Rh(I) hydride **4-142**. Then a reversible insertion of cyclohexene into the Rh–H bond gives cyclohexyl-Rh(I) intermediate **4-144**. The oxidative addition of Si–H onto Rh(I) could also give common intermediate **4-145**, which would undertake C–H reductive elimination to generate silyl-Rh(I) complex **4-146** with the release of cyclohexane. Then, the sequential C–H bond activation of the arene and C(aryl)–Si reductive elimination yield the silylarene product **4-148**.

In another example, Hartwig and co-workers reported a Rh(I)-catalyzed asymmetric intramolecular C–H bond silylation with high enantioselectivity (Scheme 4.31) [116] in the presence of norbornene as hydrogen acceptor, where the chiral bisphosphine ligands were utilized to get the high enantioselectivity. The silyl ethers involving a Si–H bond, which are formed in situ by hydrosilylation of benzophenone, undergo asymmetric C–H silylation utilizing the silyl group as a covalent directing group. They also described the detailed mechanistic investigations on this reaction in 2017 [117]. The KIE value ($k_H/k_D = 1.1$) indicated that the oxidative addition of C–H bond is not the rate-determining step. However, the KIE value ($k_H/k_D = 3.0$) from an intramolecular competition experiment suggested that the C–H bond cleavage might be the product-determining step and influences the enantioselectivity.

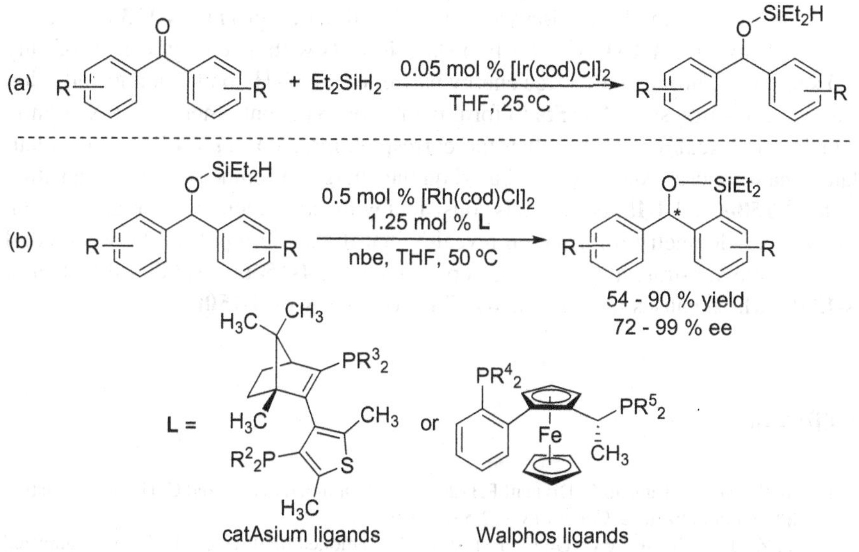

Scheme 4.31 **a** Ir(I)-catalyzed hydrosilylation reaction of benzophenones. **b** Rh(I)-catalyzed C–H bond activation and enantioselective silylation reaction of hydrido-silyl ethers

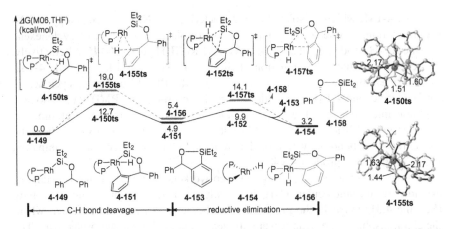

Fig. 4.32 Free energy profiles for Rh(I)-catalyzed C–H bond activation and enantioselective sily-lation reaction hydrido-silyl ethers. The values are the relative energies given in kcal/mol calculated at the M06/6-311++g(d,p)/LANL2DZ//B3-LYP/6-31g(d,p)/LANL2DZ level of theory in THF. The bond lengths are in angstroms

The calculated free energy profiles for the key steps of C–H oxidative addition and C–Si reductive elimination in Rh-catalyzed silylation are shown in Fig. 4.32. In the solid line resulting in the major (*R*)-product, the oxidation addition of C–H bond onto Rh(I) species **4-149** occurs via transition state **4-150ts** with a barrier of 12.7 kcal/mol to generate an aryl-Rh(III) hydride complex **4-151**. The following C(aryl)–Si reductive elimination yields the (*R*)-silylation product **4-153** and generates Rh(I) hydride **4-154** via transition state **4-152ts** with an energy barrier of only 5.0 kcal/mol. The calculated free energy barrier of the C–H bond oxidation addition step via transition state **4-155ts** to form the minor (*S*)-enantiomer is 19.0 kcal/mol, which is 6.3 kcal/mol higher than the corresponding step via **4-150ts**. The calculated enantiomeric excess is 99% based on the energy difference between transition states **4-150ts** and **4-155ts**, which is consistent with experimental observations of up to 99% ee. Geometry information revealed that the steric repulsion between ethyl group in the substrate and phenyl group in ligand in **4-150ts** is smaller than that in **4-155ts**, which causes the lower activation free energy of **4-150ts**.

References

1. Arockiam PB, Bruneau C, Dixneuf PH (2012) Ruthenium(II)-catalyzed C–H bond activation and functionalization. Chem Rev 112:5879–5918
2. Cai ZJ, Liu CX, Wang Q, Gu Q, You SL (2019) Thioketone-directed rhodium(I) catalyzed enantioselective C–H bond arylation of ferrocenes. Nat Commun 10:4168
3. Colby DA, Bergman RG, Ellman JA (2010) Rhodium-catalyzed C–C bond formation via heteroatom-directed C–H bond activation. Chem Rev 110:624–655

4. Gensch T, Hopkinson MN, Glorius F, Wencel-Delord J (2016) Mild metal-catalyzed C–H activation: examples and concepts. Chem Soc Rev 45:2900–2936

5. Lin C, Gao F, Shen L (2019) Advances in transition metal-catalyzed selective functionalization of inert C–O bonds assisted by directing groups. Adv Synth Catal 361:3915–3924

6. Liu CX, Gu Q, You S-L (2020) Asymmetric C–H bond functionalization of ferrocenes: new opportunities and challenges. Trends Chem 2:737–749

7. Lyons TW, Sanford MS (2010) Palladium-catalyzed ligand-directed C–H functionalization reactions. Chem Rev 110:1147–1169

8. Park Y, Kim Y, Chang S (2017) Transition metal-catalyzed C–H amination: scope, mechanism, and applications. Chem Rev 117:9247–9301

9. Song G, Li X (2015) Substrate activation strategies in rhodium(III)-catalyzed selective functionalization of arenes. Acc Chem Res 48:1007–1020

10. Song G, Wang F, Li X (2012) C–C, C–O and C–N bond formation via rhodium(iii)-catalyzed oxidative C–H activation. Chem Soc Rev 41:3651–3678

11. Sperger T, Sanhueza IA, Kalvet I, Schoenebeck F (2015) Computational studies of synthetically relevant homogeneous organometallic catalysis involving Ni, Pd, Ir, and Rh: an overview of commonly employed DFT methods and mechanistic insights. Chem Rev 115:9532–9586

12. Tauchert ME, Incarvito CD, Rheingold AL, Bergman RG, Ellman JA (2012) Mechanism of the rhodium(III)-catalyzed arylation of imines via C–H bond functionalization: inhibition by substrate. J Am Chem Soc 134:1482–1485

13. Tsai AS, Tauchert ME, Bergman RG, Ellman JA (2011) Rhodium(III)-catalyzed arylation of Boc-imines via C–H bond functionalization. J Am Chem Soc 133:1248–1250

14. Wang K, Hu F, Zhang Y, Wang J (2015) Directing group-assisted transition-metal-catalyzed vinylic C–H bond functionalization. Sci China Chem 58:1252–1265

15. Jin C, Wang G, Yang X, Zhu W, Yang Y (2018) Experimental and theoretical studies on rhodium-catalyzed direct CH benzoxylation reaction. Tetrahedron Lett 59:2042–2045

16. Hartwig JF (1998) Transition metal catalyzed synthesis of arylamines and aryl ethers from aryl halides and triflates: scope and mechanism. Angew Chem Int Ed 37:2046–2067

17. Hartwig JF (2008) Evolution of a fourth generation catalyst for the amination and thioetherification of aryl halides. Acc Chem Res 41:1534–1544

18. Surry DS, Buchwald SL (2008) Biaryl phosphane ligands in palladium-catalyzed amination. Angew Chem Int Ed 47:6338–6361

19. Figg TM, Park S, Park J, Chang S, Musaev DG (2014) Comparative investigations of Cp*-based group 9 metal-catalyzed direct C–H amination of benzamides. Organometallics 33:4076–4085

20. Ng K-H, Zhou Z, Yu W-Y (2012) Rhodium(III)-catalyzed intermolecular direct amination of aromatic C–H bonds with N-chloroamines. Org Lett 14:272–275

21. Ryu J, Shin K, Park SH, Kim JY, Chang S (2012) Rhodium-catalyzed direct C–H amination of benzamides with aryl azides: a synthetic route to diarylamines. Angew Chem Int Ed 51:9904–9908

22. Zhou B, Du J, Yang Y, Li Y (2013) Rhodium(III)-catalyzed intermolecular direct amidation of aldehyde C–H bonds with N-chloroamines at room temperature. Org Lett 15:2934–2937

23. Wang F, Yu S, Li X (2016) Transition metal-catalysed couplings between arenes and strained or reactive rings: combination of C–H activation and ring scission. Chem Soc Rev 45:6462–6477

24. Shin K, Kim H, Chang S (2015) Transition-metal-catalyzed C–N bond forming reactions using organic azides as the nitrogen source: a journey for the mild and versatile C–H amination. Acc Chem Res 48:1040–1052

25. Park J, Chang S (2015) Comparative catalytic activity of group 9 [Cp*M(III)] complexes: cobalt-catalyzed C–H amidation of arenes with dioxazolones as amidating reagents. Angew Chem Int Ed 54:14103–14107

26. Park Y, Park KT, Kim JG, Chang S (2015) Mechanistic studies on the Rh(III)-mediated amido transfer process leading to robust C–H amination with a new type of amidating reagent. J Am Chem Soc 137:4534–4542

27. Wang H, Tang G, Li X (2015) Rhodium(III)-catalyzed amidation of unactivated C(sp(3))-H bonds. Angew Chem Int Ed 54:13049–13052
28. Dateer RB, Chang S (2015) Selective cyclization of arylnitrones to indolines under external oxidant-free conditions: Dual role of Rh(III) catalyst in the C–H activation and oxygen atom transfer. J Am Chem Soc 137:4908–4911
29. Lerchen A, Knecht T, Daniliuc CG, Glorius F (2016) Unnatural amino acid synthesis enabled by the regioselective cobalt(III)-catalyzed intermolecular carboamination of alkenes. Angew Chem Int Ed 55:15166–15170
30. Yu S, Li Y, Zhou X, Wang H, Kong L, Li X (2016) Access to structurally diverse quinoline-fused heterocycles via rhodium(III)-catalyzed C–C/C–N coupling of bifunctional substrates. Org Lett 18:2812–2815
31. Yu S, Tang G, Li Y, Zhou X, Lan Y, Li X (2016) Anthranil: an aminating reagent leading to bifunctionality for both C(sp(3))-H and C(sp(2))-H under rhodium(III) catalysis. Angew Chem Int Ed 55:8696–8700
32. Liu G, Shen Y, Zhou Z, Lu X (2013) Rhodium(III)-catalyzed redox-neutral coupling of N-phenoxyacetamides and alkynes with tunable selectivity. Angew Chem Int Ed 52:6033–6037
33. Piou T, Rovis T (2015) Rhodium-catalysed syn-carboamination of alkenes via a transient directing group. Nature 527:86–90
34. Kim JY, Park SH, Ryu J, Cho SH, Kim SH, Chang S (2012) Rhodium-catalyzed intermolecular amidation of arenes with sulfonyl azides via chelation-assisted C–H bond activation. J Am Chem Soc 134:9110–9113
35. Shin K, Baek Y, Chang S (2013) Direct C–H amination of arenes with alkyl azides under rhodium catalysis. Angew Chem Int Ed 52:8031–8036
36. Park SH, Kwak J, Shin K, Ryu J, Park Y, Chang S (2014) Mechanistic studies of the rhodium-catalyzed direct C–H amination reaction using azides as the nitrogen source. J Am Chem Soc 136:2492–2502
37. Wang X, Gensch T, Lerchen A, Daniliuc CG, Glorius F (2017) Cp*Rh(III)/Bicyclic olefin cocatalyzed C–H bond amidation by intramolecular amide transfer. J Am Chem Soc 139:6506–6512
38. Xie F, Qi Z, Li X (2013) Rhodium(III)-catalyzed azidation and nitration of arenes by C–H activation. Angew Chem Int Ed 52:11862–11866
39. Wang W, Pan C, Chen F, Cheng J (2011) Copper(II)-catalyzed ortho-functionalization of 2-arylpyridines with acyl chlorides. Chem Commun 47:3978–3980
40. Chen X, Hao X-S, Goodhue CE, Yu J-Q (2006) Cu(II)-catalyzed functionalizations of aryl C–H bonds using O_2 as an oxidant. J Am Chem Soc 128:6790–6791
41. Du Z, Gao L, Lin Y (2015) Cu-catalyzed aryl C–H halogenation using N-halosuccinimides via assistance of benzoic acid. Chem Res Chin Univ 31:167–170
42. Dick AR, Hull KL, Sanford MS (2004) A highly selective catalytic method for the oxidative functionalization of C–H bonds. J Am Chem Soc 126:2300–2301
43. Bedford RB, Haddow MF, Mitchell CJ, Webster RL (2011) Mild C–H halogenation of anilides and the isolation of an unusual palladium(I)-palladium(II) species. Angew Chem Int Ed 50:5524–5527
44. Zhan B-B, Liu Y-H, Hu F, Shi B-F (2016) Nickel-catalyzed ortho-halogenation of unactivated (hetero)aryl C–H bonds with lithium halides using a removable auxiliary. Chem Commun 52:4934–4937
45. Li J-J, Mei T-S, Yu J-Q (2008) Synthesis of indolines and tetrahydroisoquinolines from arylethylamines by Pd(II)-catalyzed C–H activation reactions. Angew Chem Int Ed 47:6452–6455
46. Lied F, Lerchen A, Knecht T, Mück-Lichtenfeld C, Glorius F (2016) Versatile Cp*Rh(III)-catalyzed selective ortho-chlorination of arenes and heteroarenes. ACS Catal 6:7839–7843
47. Schroder N, Wencel-Delord J, Glorius F (2012) High-yielding, versatile, and practical [Rh(III)Cp*]-catalyzed ortho bromination and iodination of arenes. J Am Chem Soc 134:8298–8301

48. Qian G, Hong X, Liu B, Mao H, Xu B (2014) Rhodium-catalyzed regioselective C–H chlorination of 7-azaindoles using 1,2-dichloroethane. Org Lett 16:5294–5297
49. Zhang P, Hong L, Li G, Wang R (2015) Sodium halides as halogenating reagents: rhodium(III)-catalyzed versatile and practical halogenation of aryl compounds. Adv Synth Catal 357:345–349
50. Zhang T, Qi X, Liu S, Bai R, Liu C, Lan Y (2017) Computational investigation of the role played by rhodium(V) in the rhodium(III)-Catalyzed ortho-Bromination of Arenes. Chemistry 23:2690–2699
51. Matteson DS (1989) Boronic esters in stereodirected synthesis. Tetrahedron Lett 45:1859–1885
52. Suzuki A (1994) New synthetic transformations via organoboron compounds. Pure Appl Chem 66:213–222
53. Suzuki A (1991) Synthetic studies via the cross-coupling reaction of organoboron derivatives with organic halides. Pure Appl Chem 63:419–422
54. Chen H, Hartwig JF (1999) Catalytic, regiospecific end-functionalization of alkanes: rhenium-catalyzed borylation under photochemical conditions. Angew Chem Int Ed 38:3391–3393
55. Kleeberg C, Dang L, Lin Z, Marder TB (2009) A facile route to aryl boronates: room-temperature, copper-catalyzed borylation of aryl halides with alkoxy diboron reagents. Angew Chem Int Ed 48:5350–5354
56. Zhu W, Ma D (2006) Formation of arylboronates by a CuI-catalyzed coupling reaction of pinacolborane with aryl iodides at room temperature. Org Lett 8:261–263
57. Billingsley KL, Buchwald SL (2008) An improved system for the palladium-catalyzed borylation of aryl halides with pinacol borane. J Org Chem 73:5589–5591
58. Liskey CW, Hartwig JF (2013) Iridium-catalyzed C–H borylation of cyclopropanes. J Am Chem Soc 135:3375–3378
59. Molander GA, Cavalcanti LN, Garcia-Garcia C (2013) Nickel-catalyzed borylation of halides and pseudohalides with tetrahydroxydiboron [B2(OH)4]. J Org Chem 78:6427–6439
60. Chen H, Schlecht S, Semple TC, Hartwig JF (2000) Thermal, catalytic, regiospecific functionalization of alkanes. Science 287:1995–1997
61. Mkhalid IA, Barnard JH, Marder TB, Murphy JM, Hartwig JF (2010) C–H activation for the construction of C–B bonds. Chem Rev 110:890–931
62. Tobisu M, Kinuta H, Kita Y, Remond E, Chatani N (2012) Rhodium(I)-catalyzed borylation of nitriles through the cleavage of carbon-cyano bonds. J Am Chem Soc 134:115–118
63. Murphy JM, Lawrence JD, Kawamura K, Incarvito C, Hartwig JF (2006) Ruthenium-catalyzed regiospecific borylation of methyl C–H bonds. J Am Chem Soc 128:13684–13685
64. Shimada S, Batsanov AS, Howard JAK, Marder TB (2001) Formation of Aryl- and Benzyl-boronate Esters by Rhodium-Catalyzed C − H Bond Functionalization with Pinacolborane. Angew Chem Int Ed 40:2168–2171
65. Lam WH, Lam KC, Lin Z, Shimada S, Perutz RN, Marder TB (2004) Theoretical study of reaction pathways for the rhodium phosphine-catalysed borylation of C–H bonds with pinacolborane. Dalton Trans:1556–1562
66. Hartwig JF, Cook KS, Hapke M, Incarvito CD, Fan Y, Webster CE, Hall MB (2005) Rhodium boryl complexes in the catalytic, terminal functionalization of alkanes. J Am Chem Soc 127:2538–2552
67. Wei CS, Jimenez-Hoyos CA, Videa MF, Hartwig JF, Hall MB (2010) Origins of the selectivity for borylation of primary over secondary C–H bonds catalyzed by Cp*-rhodium complexes. J Am Chem Soc 132:3078–3091
68. Chang WTT, Smith RC, Regens CS, Bailey AD, Werner NS, Denmark SE, Denmark SE, Liu JHC (2011) Cross-coupling with organosilicon compounds silicon-based coupling reactions in the total synthesis of natural products. Organ React 49:213–746
69. Denmark SE, Ambrosi A (2015) Why You Really Should Consider Using Palladium-Catalyzed Cross-Coupling of Silanols and Silanolates. Org Process Res Dev 19:982–994
70. Furuta S, Mori T, Yoshigoe Y, Sekine K, Kuninobu Y (2020) Synthesis, structures and photo-physical properties of hexacoordinated organosilicon compounds with 2-(2-pyridyl)phenyl groups. Org Biomol Chem 18:3239–3242

71. Hachiya H, Hirano K, Satoh T, Miura M (2010) Nickel-catalyzed direct C–H arylation and alkenylation of heteroarenes with organosilicon reagents. Angew Chem Int Ed 49:2202–2205

72. Komiyama T, Minami Y, Hiyama T (2017) Recent advances in transition-metal-catalyzed synthetic transformations of organosilicon reagents. ACS Catal 7:631–651

73. Li C, Dou J, Cao J, Li Z, Chen W, Zhu Q, Zhu C (2013) A novel chelating organosilicone resin bearing long chain imidazolyl ligands: Preparation, characterization, and adsorption properties. J Organomet Chem 727:37–43

74. Nakao Y, Hiyama T (2011) Silicon-based cross-coupling reaction: an environmentally benign version. Chem Soc Rev 40:4893–4901

75. Prakash GKS, Yudin AK (1997) Perfluoroalkylation with organosilicon reagents. Chem Rev 97:757–786

76. Rendón-Nava D, Vásquez-Pérez JM, Sandoval-Chávez CI, Alvarez-Hernández A, Mendoza-Espinosa D (2020) Synthesis of multinuclear Rh(I) complexes bearing triazolylidenes and their application in C-C and C–Si bond forming reactions. Organometallics 39:3961–3971

77. Sore HF, Galloway WRJD, Spring DR (2012) Palladium-catalysed cross-coupling of organosilicon reagents. Chem Soc Rev 41:1845–1866

78. Cheng C, Hartwig JF (2015) Catalytic Silylation of Unactivated C–H Bonds. Chem Rev 115:8946–8975

79. Cheng C, Simmons EM, Hartwig JF (2013) Iridium-catalyzed, diastereoselective dehydrogenative silylation of terminal alkenes with (TMSO)$_2$MeSiH. Angew Chem Int Ed 52:8984–8989

80. Elsby MR, Johnson SA (2017) Nickel-catalyzed C–H silylation of arenes with vinylsilanes: rapid and reversible β-Si elimination. J Am Chem Soc 139:9401–9407

81. Esteruelas MA, Martínez A, Oliván M, Oñate E (2020) Kinetic analysis and sequencing of Si–H and C–H bond activation reactions: direct silylation of arenes catalyzed by an iridium-polyhydride. J Am Chem Soc 142:19119–19131

82. Gu Y, Shen Y, Zarate C, Martin R (2019) A Mild and direct site-selective sp2 C–H silylation of (poly)azines. J Am Chem Soc 141:127–132

83. Kakiuchi F, Igi K, Matsumoto M, Hayamizu T, Chatani N, Murai S (2002) A new chelation-assistance mode for a ruthenium-catalyzed silylation at the C–H bond in aromatic ring with hydrosilanes. Chem Lett 31:396–397

84. Kakiuchi F, Matsumoto M, Tsuchiya K, Igi K, Hayamizu T, Chatani N, Murai S (2003) The ruthenium-catalyzed silylation of aromatic C–H bonds with triethylsilane. J Organomet Chem 686:134–144

85. Kakiuchi F, Tsuchiya K, Matsumoto M, Mizushima E, Chatani N (2004) Ru3(CO)12-Catalyzed silylation of benzylic C−H bonds in arylpyridines and arylpyrazoles with hydrosilanes via C − H bond cleavage. J Am Chem Soc 126:12792–12793

86. Karmel C, Chen Z, Hartwig JF (2019) Iridium-catalyzed silylation of C–H bonds in unactivated arenes: a sterically encumbered phenanthroline ligand accelerates catalysis. J Am Chem Soc 141:7063–7072

87. Karmel C, Hartwig JF (2020) Mechanism of the iridium-catalyzed silylation of aromatic C–H bonds. J Am Chem Soc 142:10494–10505

88. Karmel C, Rubel CZ, Kharitonova EV, Hartwig JF (2020) Iridium-catalyzed silylation of five-membered heteroarenes: high sterically derived selectivity from a pyridyl-imidazoline ligand. Angew Chem Int Ed 59:6074–6081

89. Kaźmierczak J, Kuciński K, Lewandowski D, Hreczycho G (2019) Ru-Catalyzed dehydrogenative silylation of POSS-silanols with hydrosilanes: its introduction to one-pot synthesis. Inorg Chem 58:1201–1207

90. Oyamada J, Nishiura M, Hou Z (2011) Scandium-catalyzed silylation of aromatic C–H bonds. Angew Chem Int Ed 50:10720–10723

91. Richter SC, Oestreich M (2020) Emerging strategies for C–H silylation. Trends Chem 2:13–27

92. Saiki T, Nishio Y, Ishiyama T, Miyaura N (2006) Improvements of efficiency and regioselectivity in the iridium(i)-catalyzed aromatic CH silylation of arenes with fluorodisilanes. Organometallics 25:6068–6073

93. Su B, Lee T, Hartwig JF (2018) Iridium-catalyzed, β-selective C(*sp3*)-H silylation of aliphatic amines to form silapyrrolidines and 1,2-amino alcohols. J Am Chem Soc 140:18032–18038

94. Toutov AA, Liu W-B, Betz KN, Fedorov A, Stoltz BM, Grubbs RH (2015) Silylation of C–H bonds in aromatic heterocycles by an Earth-abundant metal catalyst. Nature 518:80–84

95. Wang D, Zhao Y, Yuan C, Wen J, Zhao Y, Shi Z (2019) Rhodium(II)-catalyzed dehydrogenative silylation of biaryl-type monophosphines with hydrosilanes. Angew Chem Int Ed 58:12529–12533

96. Wen J, Dong B, Zhu J, Zhao Y, Shi Z (2020) Revealing silylation of C(sp(2))/C(sp(3))-H bonds in arylphosphines by ruthenium catalysis. Angew Chem Int Ed 59:10909–10912

97. Wu Y-J, Yao Q-J, Chen H-M, Liao G, Shi BF (2020) Palladium-catalyzed ortho-C–H silylation of biaryl aldehydes using a transient directing group. Sci China Chem 63:875–880

98. Yang Y, Wang C (2015) Direct silylation reactions of inert C–H bonds via transition metal catalysis. Sci. China Chem 58:1266–1279

99. Karmel C, Li B, Hartwig JF (2018) Rhodium-catalyzed regioselective silylation of alkyl C–H bonds for the synthesis of 1,4-Diols. J Am Chem Soc 140:1460–1470

100. Lee T, Hartwig JF (2016) Rhodium-catalyzed enantioselective silylation of cyclopropyl C–H bonds. Angew Chem Int Ed 55:8723–8727

101. Zheng S, Zhang T, Maekawa H (2020) Reductive 3-silylation of benzofuran derivatives via coupling reaction with chlorotrialkylsilane. J Org Chem 85:13965–13972

102. Hua Y, Asgari P, Avullala T, Jeon J (2016) Catalytic reductive ortho-C–H silylation of phenols with traceless, versatile acetal directing groups and synthetic applications of dioxasilines. J Am Chem Soc 138:7982–7991

103. Itagaki S, Kamata K, Yamaguchi K, Mizuno N (2012) Rhodium acetate/base-catalyzed N-silylation of indole derivatives with hydrosilanes. Chem Commun 48:9269–9271

104. Lu W, Li C, Wu X, Xie X, Zhang Z (2020) [Rh(COD)Cl]₂/PPh₃-Catalyzed dehydrogenative silylation of styrene derivatives with NBE as a hydrogen acceptor. Organometallics 39:3780–3788

105. Mita T, Michigami K, Sato Y (2013) Iridium- and rhodium-catalyzed dehydrogenative silylations of C(*sp³*) H bonds adjacent to a nitrogen atom using hydrosilanes. Chem-Asian J 8:2970–2973

106. Murai M, Matsumoto K, Takeuchi Y, Takai K (2015) Rhodium-catalyzed synthesis of benzosilolometallocenes via the dehydrogenative silylation of C(*sp²*)–H bonds. Org Lett 17:3102–3105

107. Murai M, Okada R, Nishiyama A, Takai K (2016) Synthesis and properties of sila[n]helicenes via dehydrogenative silylation of C–H Bonds under rhodium catalysis. Org Lett 18:4380–4383

108. Murai M, Takeuchi Y, Yamauchi K, Kuninobu Y, Takai K (2016) Rhodium-catalyzed synthesis of chiral spiro-9-silabifluorenes by dehydrogenative silylation: mechanistic insights into the construction of tetraorganosilicon stereocenters. Chem Eur J 22:6048–6058

109. Thiot C, Wagner A, Mioskowski C (2006) Rh Soaked in Polyionic Gel: An Effective catalyst for dehydrogenative silylation of ketones. Org Lett 8:5939–5942

110. Li W, Huang X, You J (2016) Ruthenium-catalyzed intermolecular direct silylation of unreactive C(*sp³*)-H bonds. Org Lett 18:666–668

111. Liu YJ, Liu YH, Zhang ZZ, Yan SY, Chen K, Shi BF (2016) Divergent and stereoselective synthesis of beta-silyl-alpha-amino acids through palladium-catalyzed intermolecular silylation of unactivated primary and secondary C–H bonds. Angew Chem Int Ed 55:13859–13862

112. Murai M, Takami K, Takai K (2015) Iridium-catalyzed intermolecular dehydrogenative silylation of polycyclic aromatic compounds without directing groups. Chem Eur J 21:4566–4570

113. Lin Q, Lin Z, Pan M, Zheng Q, Li H, Chen X, Darcel C, Dixneuf PH, Li B (2020) Alkenes as hydrogen trappers to control the regio-selective ruthenium(ii) catalyzed ortho C–H silylation of amides and anilides. Org Chem Front. https://doi.org/10.1039/D0QO01031F

114. Ureshino T, Yoshida T, Kuninobu Y, Takai K (2010) Rhodium-catalyzed synthesis of silafluorene derivatives via cleavage of silicon-hydrogen and carbon-hydrogen bonds. J Am Chem Soc 132:14324–14326

115. Cheng C, Hartwig JF (2014) Rhodium-catalyzed intermolecular C–H silylation of arenes with high steric regiocontrol. Science 343:853
116. Lee T, Wilson TW, Berg R, Ryberg P, Hartwig JF (2015) Rhodium-catalyzed enantioselective silylation of arene C–H bonds: desymmetrization of diarylmethanols. J Am Chem Soc 137:6742–6745
117. Lee T, Hartwig JF (2017) Mechanistic studies on rhodium-catalyzed enantioselective silylation of aryl C–H Bonds. J Am Chem Soc 139:4879–4886

Printed in the United States
By Bookmasters